David Pi

Algebra of the Tetralemma

Also of Interest

Mathematical Logic
An Introduction
Daniel Cunningham, 2023
ISBN 978-3-11-078201-1, e-ISBN (PDF) 978-3-11-078207-3

Logic and Deduction
Applications to Theoretical Physics, and Number Theory
Bertrand Wong, 2026
ISBN 978-3-11-914531-2, e-ISBN (PDF) 978-3-11-221436-7

The Matrix Algebra of Logic
From Logic Operators to Cellular Automata
Eduardo Mizraji, 2026
ISBN 978-3-11-223004-6, e-ISBN (PDF) 978-3-11-223005-3

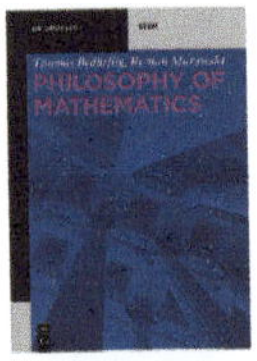

Philosophy of Mathematics
Thomas Bedürftig, Roman Murawski, 2018
ISBN 978-3-11-046830-4, e-ISBN (PDF) 978-3-11-046833-5

Nonstandard Analysis
In Higher Education, Logic and Philosophy
Karl Kuhlemann, 2025
ISBN 978-3-11-142887-1, e-ISBN (PDF) 978-3-11-142968-7

David Pi

Algebra of the Tetralemma

Truth Lattices and Logic Gates

DE GRUYTER

Mathematics Subject Classification 2020
Primary: 03B50, 06D30; Secondary: 03B52, 03G10

Author
David Pi
USA
dpi2100@gmail.com

ISBN 978-3-11-223580-5
ISBN 978-3-11-223581-2 (PDF)
ISBN 978-3-11-223582-9 (E-PUB)
DOI https://doi.org/10.1515/9783112235812

Library of Congress Control Number: 2026934888

Bibliographic information published by the Deutsche Nationalbibliothek
The Deutsche Nationalbibliothek lists this publication in the Deutsche Nationalbibliografie; detailed bibliographic data are available on the Internet at https://dnb.dnb.de.

De Gruyter and Walter de Gruyter GmbH are part of De Gruyter Brill.
www.degruyterbrill.com

Questions about General Product Safety Regulation: productsafety@degruyterbrill.com

Cover illustration: Zoonar RF / Zoonar / Getty Images Plus
Printing: CPI books GmbH, Leck

Contents

Preface —— VII

1 Introduction —— 1
1.1 Syntax —— **2**
1.2 Semantics —— **3**
1.3 Overview —— **5**

2 Nondual algebra —— 7
2.1 Lattice axioms —— **7**
2.2 Lattice order —— **8**
2.3 Involution —— **11**
2.4 Proofs to this chapter —— **15**
2.5 Exercises —— **17**

3 Measurement theory —— 18
3.1 Canonical score —— **18**
3.2 Standard score —— **19**
3.3 Negative transitivity —— **22**
3.4 Proofs to this chapter —— **25**
3.5 Exercises —— **30**

4 The tetralemma —— 31
4.1 The model —— **31**
4.2 Nominal analysis —— **33**
4.3 Four logic gates —— **36**
4.4 Proofs to this chapter —— **38**
4.5 Exercises —— **39**

5 Product algebras —— 41
5.1 Basic properties —— **42**
5.2 Measures of order —— **44**
5.3 Left–right symmetry —— **46**
5.4 Proofs to this chapter —— **50**
5.5 Exercises —— **54**

6 Fuzzy logic —— 56
6.1 Discrete fuzzy logic —— **56**
6.2 Power sets —— **58**
6.3 Fuzzy sets —— **60**

6.4 Proofs to this chapter — 63
6.5 Exercises — 64

7 Down-set representations — 66
7.1 Down-sets — 66
7.2 Up-sets — 69
7.3 Decomposition formulas — 71
7.4 Proofs to this chapter — 72
7.5 Exercises — 76

8 The middle region — 78
8.1 Convexity — 78
8.2 Kleene algebra — 80
8.3 Antichain conditions — 83
8.4 Proofs to this chapter — 87
8.5 Exercises — 92

9 Truth lattices — 94
9.1 Homomorphisms — 95
9.2 Boolean hinges — 98
9.3 Simple truth lattices — 104
9.4 Proofs to this chapter — 105
9.5 Exercises — 114

10 Algebraic logic — 116
10.1 Material implication — 117
10.2 The polarity — 122
10.3 Ockham algebras — 124
10.4 Proofs to this chapter — 129
10.5 Exercises — 134

Bibliography — 137

Index — 139

Preface

This book contains the mathematical foundations of my work on Buddhist logic. A key instrument of this system is called the tetralemma. In brief, a *tetralemma* has four premises: existence, nonexistence, both, and neither. This system goes beyond classical logic, which considers only the premises of existence and nonexistence. The possibility of "both existence and nonexistence" is eliminated by the law of excluded middle, and the possibility of "neither existence nor nonexistence" is eliminated by the dual law of noncontradiction. These laws are routinely violated in Buddhist logic.

The main thesis of the mathematical approach to Buddhist logic is that we can reason consistently in the de Morgan algebra in nondual ways that generalize the rules of the Boolean algebra. In particular, the tetralemma is a key instrument behind this agenda. We provide two basic interpretations of the four premises: One way is to model them as four truth-values in a truth lattice; and the other is to model them as four logic gates. The latter interpretation is probably closer to the informal thinking of the medieval Buddhist logicians.

This project has developed over the past few years. My efforts here represent only a humble beginning for what can be done with this theory. My earnest hope is that others will take up this project and raise it to a more mature level of discourse. I also deeply appreciate my former student, S. McClintock, for proofreading the manuscript. Since this book was written in less than a year, I must apologize if there are any remaining textual errors. However, I am confident that none of them will prove fatal.

March 30, 2026

Sincerely,
David Pi

https://doi.org/10.1515/9783112235812-203

1 Introduction

This work examines the mathematical foundations of the tetralemma within the framework of the de Morgan algebra, which is a generalization of the more familiar Boolean algebra allowing violations to the traditional laws of excluded middle and noncontradiction. This extra flexibility is useful to devise a theory of nondual logic. The tetralemma is a logical instrument originally used in Buddhist logic. In essence, it consists of four premises concerning the ontology of a designated object: it exists; it does not exist; it both exists and does not exist; it neither exists nor does not exist. In brief, we designate these four premises as existence, nonexistence, both, and neither.

The tetralemma challenges the core meaning of existence in logical analysis. The idea of "existence" is traditionally associated with the notion of truth and "nonexistence" with falsehood. In a binary world, truth and falsehood are all that matter. The *law of noncontradiction* asserts that no proposition is both true and false, and the *law of excluded middle* asserts that there is no other possibility besides true or false. Classical logic, which is the preferred logic of mathematics, is based upon this simple binary distinction. Hence, there needs to be a good reason for rejecting it. The basic rationale behind nondual logic is that the logic of duality does not adequately capture the laws of thought. The mind cannot be naively compartmentalized into existence and nonexistence, because consciousness exhibits qualities of both categories simultaneously.

The ideal logic sometimes conforms to conventional reasoning and other times it does not. Thus, we can imagine a logic that has a Boolean algebra as its nucleus (i. e., it is a subalgebra), but is flexible enough to allow some contradictions and violations to the law of excluded middle. The de Morgan algebra lends itself to this type of interpretation.

In Buddhist philosophy, the negation of both existence and nonexistence leads to "emptiness." However, this concept is not the same as the void, i. e., the empty set. Rather, it refers to the emptiness of a phenomenon. The opposite attribute is nonemptiness. There are various ways of grasping nonemptiness, but it is not the same as naïve existence. The common way of understanding existence is in terms of a permanent substance composed of irreducible units. For example, many ancient people believed that matter is composed of indestructible units which they called "atoms" or "elements." They also believed that living beings possessed an irreducible nucleus, which they called the "soul." In Buddhism, we must modify these ancient concepts of indestructible atoms and irreducible souls. The idea of nonemptiness is compatible with temporary existence and infinitely reducible substances, but it denies the existence of a mechanical world built from permanent substances.

We model the tetralemma as a special four-element de Morgan algebra. Taking the viewpoint of multivalued logic, we can interpret the four premises as four truth-values as in a truth lattice. Alternatively, we can interpret them as four logic gates. This approach provides a way to analyze the general de Morgan algebra in terms of four gates: emptiness, nonemptiness, both, and neither. Although these four gates have a certain

https://doi.org/10.1515/9783112235812-001

transcendental quality, they are also broad categories for the elements of the de Morgan algebra.

1.1 Syntax

The structure of the tetralemma, as we model it, is really easy to explain. We begin with two opposites, say "existence" and "nonexistence". Then "both" refers to their disjunction and "neither" refers to their conjunction. The diagram of the tetralemma in Figure 1.1 will appear repeatedly throughout this book:

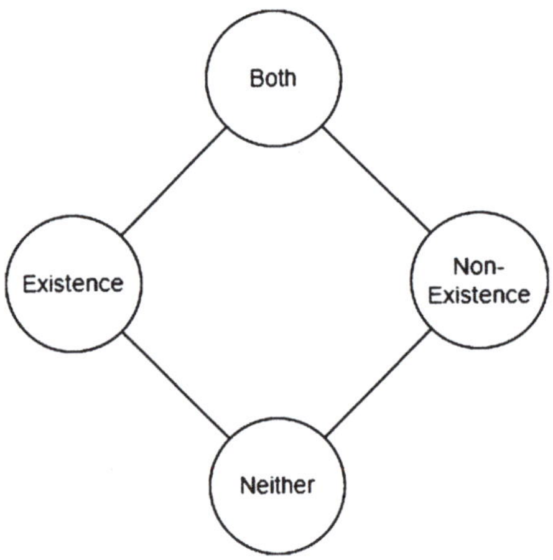

Figure 1.1: A Hasse diagram of the tetralemma.

This type of graph is called a "Hasse diagram." In essence, it is the graph of a lattice. The labels of the four nodes are secondary to the order that connects them. Of course, the notion that the conjunction of existence and nonexistence is the same as "neither existence nor nonexistence" still requires an explanation.

We are working in the framework of the de Morgan algebra, where the negation operator is called the *involution*. The involution has two defining properties. The first one is called the *law of double negation*: The involution of the involution of an object is the original object. This property is similar to the law of double negation in arithmetic: The negative of a negative number is positive. The second one is the *de Morgan laws*: the involution of a disjunction is the conjunction of the involutions; and the involution of a conjunction is the disjunction of the involutions. Applying these laws, we find that the involuted disjunction of existence and nonexistence is equivalent to the conjunction of existence and nonexistence. Thus, "both" is opposite to "neither."

In conformity with Buddhist terminology, the involuted disjunction of existence and nonexistence is defined as "emptiness," and the conjunction of existence and nonexistence refers to their "oneness." Likewise, the disjunction of existence and nonexistence refers to their "twoness." Notice that twoness is the natural opposite of oneness just as nonemptiness is the opposite of emptiness. The nodes of the tetralemma can thus be relabeled in the following suggestive manner: existence, nonexistence, emptiness, and nonemptiness. The syntax identifies "both existence and nonexistence" as nonemptiness, and "neither existence nor nonexistence" as emptiness. Figure 1.2 displays the Hasse diagram of the "tetralemma of existence."

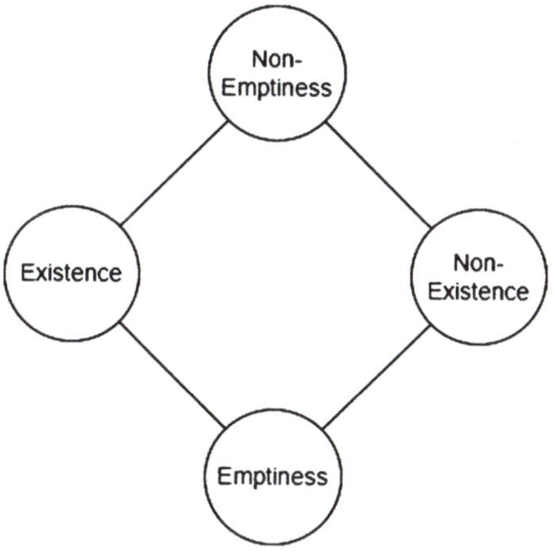

Figure 1.2: The tetralemma of existence.

We refer to the logical identity of emptiness with the oneness of existence and nonexistence as *Nagarjuna's identity*. Nagarjuna was the ancient Buddhist philosopher who originally used the tetralemma as a logical instrument. He realized that emptiness is identical to conventional existence, leading to his celebrated theory of the Middle Way. Nagarjuna's identity captures this relation formally.

1.2 Semantics

The semantics of Buddhist logic is a complex affair, and we do not intend to wander too deeply into this field. We need just enough of it to motivate some of our basic concepts, especially emptiness and nonemptiness. Taking the viewpoint of modern logic, the four premises are four truth-values arranged into the truth lattice of the tetralemma. This is the formal semantics of multivalued logic. Although this approach is useful, it is not the only way that we apply the logic of the tetralemma.

The second basic interpretation of the tetralemma is in terms of four logic gates for the general de Morgan algebra X. We begin with the premises of existence and nonexistence, because the associated gates for them are already familiar ones. Indeed, we associate "existence" with the identity map that takes each object into itself, and "nonexistence" with the involution operator that takes each object into its negation. The concept of negation that is being used here is not merely the absence of existence. Rather, it indicates the replacement of an object with another object that is its "involution." The involution is ordinarily a distinct object, which is the "opposite" of the object that is involuted. However, an object can also be self-negated, i. e., the involuted object is the same as the original object. Of course, we are all very familiar with a self-negated object in arithmetic, i. e., the number zero. It is actually a relatively difficult number to grasp, because it is intangible.

The transcendental gates of emptiness and nonemptiness are even more difficult to conceptualize. We associate "emptiness" with the map that takes an object into the conjunction of itself with its involution, and we associate "nonemptiness" with the map that takes an object into the disjunction of itself with its involution. But what kind of object is empty (or nonempty) in this sense? In Buddhist parlance, every phenomenon is both empty and nonempty. However, we are devising a formal language that assigns the attribute of emptiness to some objects and nonemptiness to others. For example, these objects can be propositions in a formal logic. We define an object as *empty* if it is within the codomain of the gate of emptiness, and we define an object as *nonempty* if it is within the codomain of the gate of nonemptiness. The ontological concepts are thus redefined as algebraic forms: An object is empty provided that it is the conjunction of some object (x) with its involution (not x); and an object is nonempty provided that it is the disjunction of some object (x) with its involution (not x).

We thus arrive at four transcendental categories: empty-only, nonempty-only, both, and neither. The category of "both" refers to those objects that are both empty and nonempty. The only objects like this are self-negated objects. The category of "neither" refers to those objects that are neither empty nor nonempty. For example, we can classify the four truth-values of a tetralemma with this system: The truth-value "both" is nonempty; the truth-value "neither" is empty; the truth-value "existence" is neither; and the truth-value "nonexistence" is neither. One must look past the labels and grasp the deeper meaning as their functional relations.

Western logic is preoccupied with the preservation of truth. That approach is certainly essential in mathematics, but it is a complex task in nondual logic. Perhaps the most well-known effort in this direction is the Heyting algebra, which develops a novel concept of implication. This algebra embraces a concept of negation that violates the law of double negation, i. e., the pseudocomplement. We think that this is a huge sacrifice. On the other hand, the de Morgan algebra accepts the law of double negation, but it does not seem to shed enough light upon the paradoxes of material implication. Yet, there is a concept of entailment available in the lattice order. Moreover, polarity acts like a "contrapositive" that reverses the direction of the lattice order. We will contemplate alternatives to the de Morgan algebra in the last chapter, but our results only reinforce the role of the de Morgan algebra as the key vehicle for algebraic logic.

1.3 Overview

This book is designed as an advanced undergraduate textbook. Mathematical proofs of the main results are found in the appendices at the end of each chapter. This feature allows students to concentrate on basic concepts before plunging into the details of mathematical confirmation. Moreover, each of the following chapters is equipped with a set of exercises designed for beginners in abstract mathematical reasoning. This book relies heavily upon Hasse diagrams to illustrate concepts through examples and counterexamples. We provide some "alternate text" at the end of the book to briefly describe the contents of each figure for the sake of the visually impaired. We next outline what this book intends to accomplish, beginning with the next chapter.

Chapter 2 provides the basic concepts of the de Morgan algebra. We review the two basic ways of axiomatizing lattice theory in terms of logical equations or the lattice order. The analysis is light. We discuss the equivalence of the law of opposites and the law of double negation, and the relation between the law of polarity and the de Morgan laws for involution. This analysis integrates the involution into the ordinal structure of the underlying lattice. Finally, we present Nagarjuna's identity. This simple result is trivial in the Boolean algebra, but it plays a major role in the algebra of the tetralemma.

Chapter 3 deals with measurement theory. We explore two methods for one-way measuring the lattice order of a finite algebra: the canonical score and the standard score. The former is more general, because the standard score is defined only for a modular lattice. However, the latter is simpler, because it is based only upon ranks. The necessary and sufficient condition for strengthening these scores to two-way measurement is the same: The strict order of the lattice must be negatively transitive.

Chapter 4 introduces the tetralemma. We construct the four premises as both truth-values and logic gates. We examine various algebras with four elements and discuss why the design we have chosen is the best design of the tetralemma. The analysis of the four premises as four logic gates takes the theory to a new level. We define the emptiness and nonemptiness of objects in the de Morgan algebra as special forms, and investigate their basic properties. In brief, empty objects other than the bottom violate the law of noncontradiction, and nonempty objects other than the top violate the law of excluded middle. Thus, the de Morgan algebra is non-Boolean if and only if these types of objects are present.

Chapter 5 investigates the structure of product algebras. In T'ien-tai philosophy, the product signifies "mutual inclusion," i. e., each object from one factor "possesses" all the objects in the other factor. Product algebras inherit various algebraic properties from their constituent factors.

Chapter 6 examines the structure of fuzzy logic. This is a special type of de Morgan algebra characterized by the totality of the lattice order, i. e., every pair of objects is comparable with respect to the lattice order. We apply the theory of product algebras to extend the logic to the theory of fuzzy sets.

Chapter 7 provides a representation scheme of antichains with down- or up-sets. In particular, each element in the de Morgan algebra is associated with a principal down- or up-set. This theory is a modest generalization of ideal or filter theory. It is especially interesting, because the set of all empty objects is a down-set, and the set of all nonempty elements is an up-set.

Chapter 8 investigates the "middle region" of the de Morgan algebra. This region consists of all self-negated objects and all pairs of incomparable opposites. These are the objects that belong to the transcendental categories of both and neither. The middle region is a convex set. As a special case, we analyze the middle region of the Kleene algebra. Furthermore, we provide necessary and sufficient conditions for the middle region of the product algebra to be an antichain.

Chapter 9 examines the theory of truth lattices for de Morgan algebras. A truth lattice is a useful device for assigning truth-values to propositions only if there is a homomorphism from the algebra of propositions onto the truth lattice. We investigate the scope of the Boolean truth lattice, the logic of paradox, and the tetralemma as truth lattices. The theory of truth lattices for Boolean algebras is an important special case.

Chapter 10 examines the theory of implication from the viewpoint of algebraic logic. The two main types of inference involve material implication and lattice order. A key issue is the status of the reverse implication, e. g., the contrapositive. When inferences are made with the lattice order, we must scrutinize the law of polarity. This agenda becomes more difficult if we do not rely upon the law of double negation.

2 Nondual algebra

The type of algebra that we have in mind involves three logical operators: disjunction, conjunction, and involution. The theory is fully developed here only for a finite algebra, but many results obviously generalize to the infinite case. The basic axioms of the theory are universal definitions regardless of the number of objects in a model.

We also freely use the language of universal algebra. One idea that we borrow is the *subalgebra*, i. e., a subset of objects from the algebra for which the operations are closed. Subalgebras reveal hidden relations among the objects in the algebra. We also borrow the language of morphisms (or functors) mapping one algebra into another. By definition, the morphism $f : X \to Y$ is *injective* or *one-to one* provided that $f(x_1) = f(x_2)$ implies $x_1 = x_2$. It is *surjective* or *onto* provided that $Y = \{f(x) : x \in X\}$. The morphism f is *invertible* if and only if is both injective and surjective.

The most basic morphism is the *homomorphism*, i. e., a functor from an algebra into another of the same type that preserves the logic operators. In particular, the *monomorphism* is a one-to-one homomorphism, and the *isomorphism* is an onto monomorphism. Isomorphic algebras are mathematically the same. As a special case, the *automorphism* is an isomorphism from the algebra onto itself. The presence of an automorphism indicates a deep symmetry in the structure of the algebra.

There are many types of algebras that have been developed as alternatives to the Boolean algebra. In particular, the Ockham algebra is a very general class that includes the Stone and Heyting algebras as special cases. However, we intend to focus on the *de Morgan algebra*, which is a relatively simple but underutilized logical system. One possible definition of the de Morgan algebra is that it is an Ockham algebra that satisfies the law of double negation. This algebra is especially attractive, because of the elegant way it handles the Buddhist concept of "emptiness." This topic will be developed mathematically throughout this work.

2.1 Lattice axioms

We first review the basic algebraic structure of a *lattice*. The classic reference to this material is the work of Birkhoff [1], but a more modern treatment is given by Grätzer [2]. This structure has two binary operators called the conjunction and disjunction. Let X denote a collection of objects with generic objects x and y. The notation $x \wedge y$ represents a third object in X called the "conjunction of x and y." In formal terms, the *conjunction* is a binary operator $\wedge : X^2 \to X$ defined by the following four axioms:

The conjunction

(1) Idempotency: $x \wedge x = x$;
(2) Commutativity: $x \wedge y = y \wedge x$;
(3) Associativity: $(x \wedge y) \wedge z = x \wedge (y \wedge z)$;
(4) Absorption: $x \wedge (x \vee y) = x$.

https://doi.org/10.1515/9783112235812-002

Likewise, the notation $x \vee y$ represents the "disjunction of x and y." In formal terms, the *disjunction* is a binary operator $\wedge : X^2 \to X$ defined by the following four axioms:

The disjunction

(1) Idempotency: $x \vee x = x$;
(2) Commutativity: $x \vee y = y \vee x$;
(3) Associativity: $(x \vee y) \vee z = x \vee (y \vee z)$;
(4) Absorption: $x \vee (x \wedge y) = x$.

At this level of discourse, the key distinction between disjunction and conjunction is the absorption laws, which explain how conjunction and disjunction interact. Notice that the interchange of the lattice symbols $\wedge$ and $\vee$ transforms the axioms for conjunction into the axioms for disjunction, and vice versa. This fact gives rise to the most basic lattice duality: The interchange of $\wedge$ and $\vee$ in any theorem about lattices leads to another (dual) theorem about lattices.

There are two auxiliary assumptions about lattices that are commonly evoked. The first one is a pair of *distributive laws*:

$$x \wedge (y \vee z) = (x \wedge y) \vee (x \wedge z);$$
$$x \vee (y \wedge z) = (x \vee y) \wedge (x \vee z).$$

In rough terms, the former law asserts that conjunction distributes over disjunction, and the latter asserts that disjunction distributes over conjunction. This assumption is important in Birkhoff's representation theory of lattices [3], but it has only secondary interest here. We will evoke it occasionally.

The second auxiliary assumption is the *modular law*:

$$(a \wedge b) \vee (x \wedge b) = \left[(a \wedge b) \vee x\right] \wedge b.$$

This law is a weaker condition than the distributive laws. It is somewhat more useful to our endeavor, mainly because of the convenient way it simplifies the lattice structure.

2.2 Lattice order

The lattice structure is deeply connected to a logical order that it imposes upon the objects. By definition, a *partial order* on X is a binary relation $\succcurlyeq$ that satisfies the following three properties:

(1) Reflexivity: $x \succcurlyeq x$;
(2) Antisymmetry: $x \succcurlyeq y$ and $y \succcurlyeq x$ implies $x = y$;
(3) Transitivity: $x \succcurlyeq y$ and $y \succcurlyeq z$ implies $x \succcurlyeq z$.

It is sometimes convenient to write the partial order in the reverse direction. By convention, $y \preccurlyeq x$ is equivalent to $x \succcurlyeq y$. The partial order $\succcurlyeq$ can be decomposed into its symmetric and asymmetric parts. The *asymmetric component* $\succ$ is defined by $x \succ y$ if and only if $x \succcurlyeq y$ and not $y \succcurlyeq x$. In the reverse direction, $y \prec x$ is equivalent to $x \succ y$. This relation has two defining properties:

(1) Asymmetry: $x \succ y$ implies not $y \succ x$;
(2) Transitivity: $x \succ y$ and $y \succ z$ implies $x \succ z$.

Informally, we say that "x is greater than y" or "x is above y" provided that $x \succ y$. On the other hand, the *symmetric component* is the equivalence relation $=$ defined by $x = y$ if and only if $x \succcurlyeq y$ and $y \succcurlyeq x$. As an equivalence relation, it satisfies the following three properties:

(1) Reflexivity: $x = x$;
(2) Symmetry: $x = y$ implies $y = x$;
(3) Transitivity: $x = y$ and $y = z$ implies $x = z$.

We say that the relation $\succcurlyeq$ is *total* provided that either $x \succcurlyeq y$ or $y \succcurlyeq x$ for every pair x and y. The adjective "partial" indicates that the relation is not necessarily total. The *incomparability relation* $\parallel$ is defined by the condition that $x \parallel y$ if and only if not $x \succcurlyeq y$ and not $y \succcurlyeq x$. For every pair of objects x and y, there are only these four possibilities: (i) $x \succcurlyeq y$; (ii) $y \succcurlyeq x$; (iii) "both," i. e., $x = y$; and (iv) "neither," i. e., $x \parallel y$. Combining "both" and "neither," we define the relation $\simeq$ by the condition that $x \simeq y$ if and only if $x = y$ or $x \parallel y$.

Suppose X is a lattice. Then the *lattice order* $\succcurlyeq$ on X is defined by the condition that $x \succcurlyeq y$ if and only if $x = x \vee y$. It follows from duality that $x = x \vee y$ if and only if $y = x \wedge y$. Furthermore, the following basic identities hold:

$$x \vee y = \inf\{z : z \succcurlyeq x, z \succcurlyeq y\};$$
$$x \wedge y = \sup\{z : x \succcurlyeq z, y \succcurlyeq z\}.$$

Conversely, one can begin with these equations as the definition of disjunction and conjunction, and construct the lattice axiomatically from a partial order. The key assumption is that these equations are well defined, i. e., the infimum and the supremum must exist uniquely. Then one must verify that the lattice axioms are satisfied.

An object $\top$ is a *top* provided that $\top \succcurlyeq x$ for every object $x \in X$. Similarly, an object $\bot$ is a *bottom* provided that $x \succcurlyeq \bot$ for every object $x \in X$. A lattice with both a top and bottom is said to be *bounded*. In particular, a finite lattice is always bounded with top $\top = \bigvee_{x \in X} x$ and bottom $\bot = \bigwedge_{x \in X} x$.

The lattice order of a finite lattice can be displayed as a type of directed graph known as a Hasse diagram. By definition, the object x is a *successor* to object y whenever $x \succ y$ and there is no intermediate object z such that $x \succ z \succ y$; and we say that the object y is a *predecessor* to object x provided that object x is a successor to object y. The *Hasse diagram* is a directed graph consisting of nodes representing the objects and vertical or near vertical lines connecting nodes to their successors and predecessors. A (finite) *chain* from y to x is a path from predecessors to successors that begins with y and ends with x. Thus, the chain has the form $x = x_m \succ \cdots \succ x_2 \succ x_1 = y$, where x_{i+1} is a successor to x_i. In a finite lattice, $x \succ y$ if and only if there is a chain from y to x.

Example. Figure 2.1 displays the Hasse diagram for a "diamond lattice":

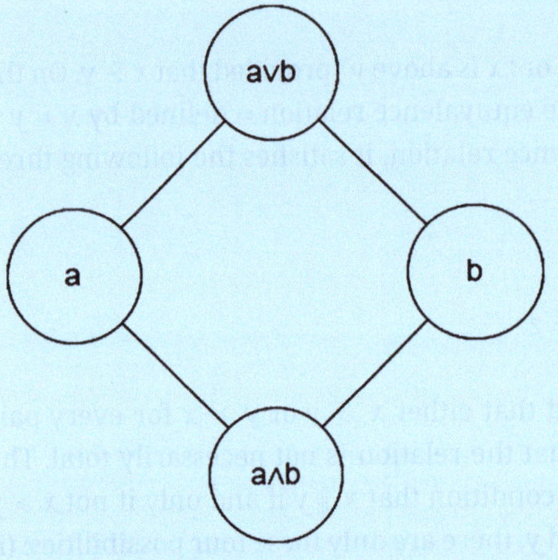

Figure 2.1: The diamond lattice.

The diamond lattice does not evoke any concept of negation, but it is the prototype of a tetralemma. The two objects a and b are incomparable. Their disjunction is literally above them and their conjunction is literally below them. There are two distinct chains from $a \wedge b$ to $a \vee b$. If we combine any two objects with lattice operations, the result reduces to one of the above four objects. For example, one can verify from the algebraic laws of syntax that $(a \vee b) \vee (a \wedge b) = a \vee b$. The Hasse diagram makes this identity transparent, because $a \vee b \succ a \wedge b$.

For any $y \succcurlyeq z$, define the *interval* from z to y by $[z, y] := \{x \in X : y \succcurlyeq x \succcurlyeq z\}$. Referring to Figure 2.1, the diamond lattice consists of four intervals:

$$[a \wedge b, a], \quad [a \wedge b, b], \quad [a, a \vee b], \quad \text{and} \quad [b; a \vee b].$$

Moreover, they are arranged so that $[a \wedge b, a]$ is on the opposite side of the diamond to $[b, a \vee b]$. Likewise, $[a \wedge b, b]$ is on the opposite side to $[a, a \vee b]$. The lattice order sometimes provides an elegant way to envision some deep results in lattice theory. In particular, the well-known *Dedekind criterion* [4] asserts that a lattice is modular if and only if it does not contain a sublattice isomorphic to the "pentagonal lattice." This lattice consists of five elements arranged in two chains from bottom to top. One chain is necessarily longer than the other. Figure 2.2 provides the Hasse diagram for this lattice.

The lattice order is incorporated into the following alternative condition for the modularity law:

$$y \succcurlyeq z \text{ implies } (y \wedge x) \vee z = y \wedge (x \vee z).$$

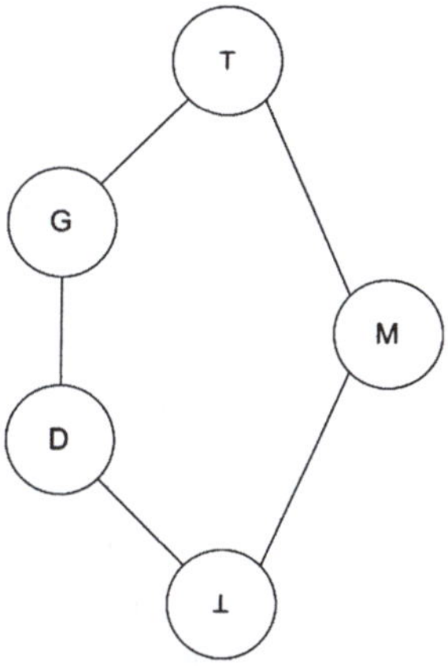

Figure 2.2: The Hasse diagram of a pentagonal lattice.

This equivalent formula is often easier to apply than the original definition. In particular, we can reanalyze the diamond lattice in Figure 2.1 with it. Suppose it is a sublattice of a modular lattice X. It is easy to verify that the well-defined map $f : [b, a \vee b] \to [a \wedge b, a]$ given by $f(x) = x \wedge a$ is order-preserving. Likewise, the well-defined map $g : [a \wedge b, a] \to [b, a \vee b]$ given by $g(y) = y \vee b$ is order-preserving. The modular law guarantees that these two maps are inverses. Hence, we arrive at the *diamond isomorphism theorem*: The opposites sides of the diamond lattice are order isomorphic.

2.3 Involution

The third operator that we need is unary. By definition, the *involution* $\sim$ is a map $\sim : X \to X$ that satisfies the following two laws:

(1) Double negation: $\sim(\sim x) = x$;
(2) Polarity: $x \succ y$ if and only if $\sim y \succ \sim x$.

The law of double negation imposes symmetry upon the algebra. It is easy to verify that it is equivalent to the *law of opposites*: $y = \sim x$ if and only if $x = \sim y$. This law was regarded as common knowledge among many ancient philosophers. For example, it corresponds to the yin-yang theory in Taoism – every yin has a matching yang.

Lemma 2.1. *The law of double negation is equivalent to the law of opposites.*

We regard x and $\sim x$ as genuine opposites as long as $x \neq \sim x$. On the other hand, the relation $x = \sim x$ is the enigma of a *self-negating object*.

Suppose the law of double negation holds. Then it is obvious that the law of polarity is equivalent to the condition that $x \succcurlyeq y$ if and only if $\sim y \succcurlyeq \sim x$. This law is also deeply connected to the *de Morgan laws*:

$$\sim(x \vee y) = \sim x \wedge \sim y;$$
$$\sim(x \wedge y) = \sim x \vee \sim y.$$

These laws verify that the involution map $\sim : X \to X$ is a *dual automorphism*, i. e., it is invertible and exchanges conjunctions and disjunctions.

Lemma 2.2. *Suppose X is a lattice and the unary operator $\sim : X \to X$ satisfies the law of double negation. Then the operator $\sim$ satisfies the law of polarity if and only if it satisfies the de Morgan laws.*

These two laws are named in honor of Augustus de Morgan, but they have been traced further back to William Ockham [5]. The two laws are equivalent provided that the law of double negation holds. In view of Lemma 2.1, an alternative definition of an involution is a unary operator that satisfies the law of double negation and the de Morgan laws. In fact, this is the most common definition.

The law of polarity fits naturally into Buddhist logic. "Reason" in this context means to explain the causes and conditions. Suppose we interpret the lattice order as a causal implication. Thus, $x \succ y$ means that "y is the cause" and "x is the effect." In brief, "cause implies effect." This causal relation makes sense in Buddhist logic provided that $\sim y \succ \sim x$, i. e., "no effect implies no cause."

The *de Morgan algebra* is a bounded lattice equipped with an involution. This is the primary algebraic structure that we investigate in this work. The originator of the de Morgan algebra was Grigore Moisil [6], who named it after Augustus de Morgan. However, Moisil did not require the lattice to be bounded. We are also breaking with tradition by not requiring the lattice to be distributive. If this happens to be the case, we simply refer to the structure as a *distributive de Morgan algebra*. Likewise, a *modular de Morgan algebra* is a de Morgan algebra such that the underlying lattice is modular.

The *Boolean algebra* is a distributive de Morgan algebra that satisfies the following pair of auxiliary laws:
(1) Law of excluded middle: $x \vee \sim x = \top$;
(2) Law of noncontradiction: $x \wedge \sim x = \bot$.

These laws originally belong to Aristotle's Organon. However, the Boolean algebra is the modern foundation of conventional logic. A basic way to build a nonconventional logic is to loosen the restrictions of the Boolean algebra. There are many ways to do this, but the de Morgan algebra is perhaps the most elegant model of a nondual algebra. In this context, "nonduality" refers to an algebra that violates one or both of the above pair of laws.

For a key case study, let us briefly examine the logic of sets. Suppose the universe Ω is a collection of elements, and the objects are subsets of Ω. The collection of all objects is thus the power set 2^Ω. The subsets are combined with three basic set-theoretic operators: union, intersection, and complementation. They have well-known definitions. Let α and β denote two subsets of Ω. The *union* $\alpha \cup \beta$ is the subset defined by $\alpha \cup \beta := \{x \in \Omega : x \in$

a or $x \in \beta\}$; and the *intersection* $a \cap \beta$ is the subset defined by $a \cap \beta := \{x \in \Omega : x \in a$ and $x \in \beta\}$. The *complement* a' of the subset a is the subset defined by $a' := \{x \in \Omega : x \notin a\}$.

It is well known that this structure is a Boolean algebra such that disjunction is union, conjunction is intersection, and involution is complementation. The lattice order is given by set inclusion. Furthermore, the top is given by $\top = \Omega$ and the bottom is given by $\bot = \emptyset$, the empty set. In this context, the law of excluded middle asserts that

$$a \cup a' = \Omega,$$

and the law of noncontradiction asserts that

$$a \cap a' = \emptyset.$$

In words, the former law asserts that a subset and its complement collectively exhaust the universe, and the latter law asserts that a subset and its complement are mutually exclusive. The logic of sets is behind Aristotle's syllogistic reasoning. We must abandon this way of thinking as we analyze the de Morgan algebra.

Another way to loosen the axioms of the Boolean algebra is to drop the condition that the algebra is distributive. A de Morgan algebra that satisfies the laws of excluded middle and noncontradiction is called an *ortholattice*.

Example. The de Morgan algebra X displayed in Figure 2.3 is an ortholattice.

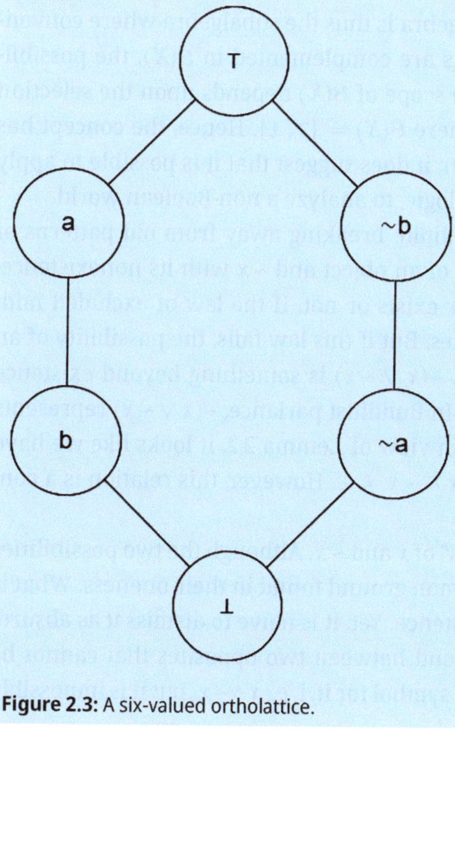

Figure 2.3: A six-valued ortholattice.

Notice that the subset $\{\top, a, b, \sim b, \bot\}$ is a pentagonal sublattice of X. The Dedekind criterion thus indicates that X is nonmodular and, hence, nondistributive. Therefore, X is a non-Boolean algebra.

Lemma 2.3. *The logical identity* $\sim(x \vee \sim x) = x \wedge \sim x$ *holds for the de Morgan algebra.*

We refer to this formula as *Nagarjuna's identity* in honor of the great Buddhist metaphysician Nagarjuna. We say that an element $x \in X$ is *complemented* provided that $x \vee \sim x = \top$. This condition is equivalent to $x \wedge \sim x = \bot$ by virtue of Nagarjuna's identity. All elements are complemented in an ortholattice. Therefore, Nagarjuna's identity is trivial in this case, because it reduces to $\sim \top = \bot$. More broadly, this identity indicates that an object in a non-Boolean algebra violates the law of excluded middle if and only if it violates the law of contradiction. However, there are other ways to interpret it.

The *Boolean center* of the de Morgan algebra X is defined by

$$\mathcal{B}(X) := \{x \in X : x \vee \sim x = \top\}.$$

The basic idea of the next result is well known, but we have not found it presented in quite this form.

Theorem 2.4. *If X is a distributive de Morgan algebra, then the Boolean center $\mathcal{B}(X)$ is a Boolean subalgebra of X.*

The Boolean center of the de Morgan algebra is thus the subalgebra where conventional reasoning prevails. Since all elements are complemented in $\mathcal{B}(X)$, the possibility of contradictions is eliminated here. The scope of $\mathcal{B}(X)$ depends upon the selection of X. We will encounter many examples where $\mathcal{B}(X) = \{\top, \bot\}$. Hence, the concept has no direct relevance in our analysis. However, it does suggest that it is possible to apply mathematics, which is based upon Boolean logic, to analyze a non-Boolean world.

We need some new terminology to facilitate breaking away from old patterns of thought. If we identify x with the existence of an object and $\sim x$ with its nonexistence, then $x \vee \sim x$ indicates that the object either exists or not. If the law of excluded middle holds, then these are the only possibilities. But if this law fails, the possibility of an "intermediate existence" emerges. In brief, $\sim(x \vee \sim x)$ is something beyond existence and nonexistence, something other than $\bot$. In Buddhist parlance, $\sim(x \vee \sim x)$ represents "emptiness," but this is not the empty set. In view of Lemma 2.2, it looks like we have reached a "logical contradiction," because $x \wedge \sim x \neq \bot$. However, this relation is a contradiction only in the old way of thinking.

We now interpret $x \wedge \sim x$ as the "oneness" of x and $\sim x$. Although the two possibilities of x and $\sim x$ are opposites, they have a common ground found in their oneness. What is oneness? It is neither existence nor nonexistence. Yet, it is naïve to dismiss it as absurd. The concept of oneness refers to a deep bond between two opposites that cannot be explained in conventional terms. We have a symbol for it, i. e., $x \wedge \sim x$, but it is impossible to explain in words the source of this unity.

Let us place this reasoning in the format of a diamond lattice (see Figure 2.4). Although this configuration is a self-contained de Morgan algebra, we can also envision it as the subalgebra of a greater de Morgan algebra for which x and $\sim x$ are incomparable opposites. The top is the disjunction of this pair of opposites, which represents their "twoness"; and the bottom is their conjunction, which represents their "oneness." In view of Nagarjuna's identity, oneness and twoness are also opposites. Thus, we interpret $x \vee \sim x$ as the "nonemptiness" of the object. However, nonemptiness and emptiness are comparable opposites in this algebra, i. e., $x \vee \sim x > x \wedge \sim x$. This structure is called a *tetralemma*, and this is our first rigorous exposure to it.

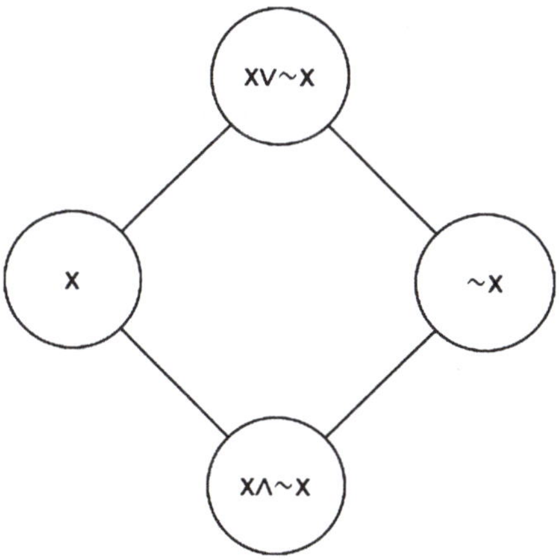

Figure 2.4: The algebraic form of a tetralemma.

The way that Nagarjuna defined emptiness involves the entire tetralemma framed as four premises: existence, nonexistence, both, and neither. He often presented an argument that rejected all four premises, concluding that the object is empty. However, it is easy to verify that this procedure reduces to the bottom of the tetralemma.

Lemma 2.5. *The involuted disjunction of* $\{x, \sim x, x \vee \sim x,\ x \wedge \sim x\}$ *is equivalent to* $x \wedge \sim x$.

The conclusion is quite profound. As long as we remain within the logic of the de Morgan algebra, Nagarjuna's definition of emptiness and the simpler definition presented here are equivalent.

2.4 Proofs to this chapter

Lemma 2.1. *The law of double negation is equivalent to the law of opposites.*

Proof. First, assume that the law of double negation holds. Then $p = \sim q$ obviously implies $\sim p = \sim\sim q = q$. Likewise, $q = \sim p$ implies $\sim q = \sim\sim p = p$. Therefore, the law of opposites holds.

Conversely, assume that the law of opposites holds. Let $q = \sim p$. Then we directly obtain $\sim q = \sim\sim p$. The law of opposites yields $p = \sim q$. We thus obtain $p = \sim\sim p$. Therefore, the law of double negation holds. □

Lemma 2.2. *Suppose X is a lattice and the unary operator $\sim : X \to X$ satisfies the law of double negation. Then the operator $\sim$ satisfies the law of polarity if and only if it satisfies the de Morgan laws.*

Proof. First, assume that the de Morgan laws hold. Then we obtain $a = a \vee b$ implies $\sim a = \sim a \wedge \sim b$. Thus, $a \succcurlyeq b$ implies $\sim b \succcurlyeq \sim a$. Conversely, $\sim b \succcurlyeq \sim a$ implies $\sim\sim a \succcurlyeq \sim\sim b$. Applying the law of double negation, we obtain $a \succcurlyeq b$. We thus conclude $a \succcurlyeq b$ if and only if $\sim b \succcurlyeq \sim a$. Next, $a \succ b$ (i. e., $a \succcurlyeq b$ and $b \not\succcurlyeq a$) if and only if $\sim b \succ \sim a$, which means that the law of polarity holds.

Conversely, assume that the law of polarity holds. By definition, $a \vee b = \inf\{U(a) \cap U(b)\}$ and $\sim a \wedge \sim b = \sup\{D(\sim a) \cap D(\sim b)\}$. Moreover, the law of polarity yields $\sim U(a) = D(\sim a)$ and $\sim U(b) = D(\sim b)$. Therefore, we obtain

$$\sim(a \vee b) = \sup\{\sim U(a) \cap \sim U(b)\} = \sup\{D(\sim a) \cap D(\sim b)\} = \sim a \wedge \sim b.$$

The proof of the other de Morgan law can be constructed similarly. But one de Morgan law and double negation also entails the other de Morgan law. □

Lemma 2.3. *The logical identity $\sim(x \vee \sim x) = x \wedge \sim x$ holds for the de Morgan algebra.*

Proof. First, an application of de Morgan's law yields $\sim(x \vee \sim x) = \sim x \wedge \sim\sim x$. Next the law of double negation yields $\sim x \wedge \sim\sim x = \sim x \wedge x$. □

Theorem 2.4. *If X is a distributive de Morgan algebra, then the center $\mathcal{B}(X)$ is a Boolean subalgebra of X.*

Proof. Notice the top $\top$ and bottom $\bot$ are complemented. Thus, the set of all complemented elements is a nonempty set. It is also clear from the law of double negation that an element x is complemented if and only if it involution $\sim x$ is complemented. We next show that the set of complemented elements is closed with respect to disjunction and conjunction.

First, we show closure with respect to disjunction. Assume two elements, say x and y, are complemented, i. e., $x \vee \sim x = \top$ and $y \vee \sim y = (x \vee y) \vee \sim(x \vee y)$. Thus, we want to show that $x \vee y$ is complemented. We compute

$$(x \vee y) \vee \sim(x \vee y) = (x \vee y) \vee (\sim x \wedge \sim y) = x \vee [y \vee (\sim x \wedge \sim y)].$$

Working within the brackets,

$$y \vee (\sim x \wedge \sim y) = (y \vee \sim x) \wedge (y \vee \sim y)$$

by virtue of the distributive law. Since $y \vee \sim y = \top$, the above expression reduces to $y \vee \sim x$. Therefore, we obtain

$$(x \vee y) \vee \sim(x \vee y) = x \vee (y \vee \sim x) = (x \vee \sim x) \vee y.$$

Since $x \vee \sim x = \top$, the above expression reduces to $\top$. Therefore, $x \vee y$ is complemented.

We next verify that $x \wedge y$ is complemented. Notice that $\sim(x \wedge y) = \sim x \vee \sim y$. Moreover, $\sim x$ is complemented, because x is complemented; and $\sim y$ is complemented, because y is complemented. Therefore, $\sim x \vee \sim y$ is complemented, which implies $\sim(\sim x \vee \sim y) = x \wedge y$ is complemented.

We have now verified that the set of all complemented elements is a subalgebra of X. It is a Boolean subalgebra, because its top is $\top$ and its bottom is $\bot$. □

Lemma 2.5. *The involuted disjunction of* $\{x, \sim x, x \vee \sim x,\ x \wedge \sim x\}$ *is equivalent to* $x \wedge \sim x$.

Proof. Let $a = x \vee \sim x$. Then the four terms are represented by $\{x, \sim x, a, \sim a\}$. Thus, their disjunction is given by

$$(x \vee \sim x) \vee a \vee \sim a = a \vee a \vee \sim a = a \vee \sim a.$$

Since $a \succcurlyeq \sim a$, we obtain $a \vee \sim a = a$. Therefore, the negated disjunction is given by $\sim a = x \wedge \sim x$. □

2.5 Exercises

(1) Construct a partial order of three elements that is not a lattice order.

(2) Suppose the unary operator $f : X \to X$ on the lattice X satisfies the law of double negation. Demonstrate that the two de Morgan laws are equivalent.

(3) Suppose the unary operator $f : X \to X$ onto the lattice X satisfies the two de Morgan laws, but not necessarily the law of double negation. Demonstrate that the map $f^2 := f \circ f$ is a lattice automorphism.

(4) Demonstrate that a de Morgan algebra with five distinct elements has at least one self-negated element.

(5) Construct a de Morgan algebra with five distinct elements such that three pairs are incomparable.

(6) Demonstrate that there is no distributive de Morgan algebra with five distinct elements such that three pairs are incomparable.

(7) Demonstrate that there does not exist a de Morgan algebra with five distinct elements such that exactly one pair of them are incomparable.

(8) Suppose X is a de Morgan algebra. Demonstrate that the distributive law implies the modular law.

(9) Verify that the pentagonal lattice violates the modularity law.

(10) Suppose X is a modular lattice. Verify that the diamond isomorphism theorem holds.

3 Measurement theory

We next study logical order from the viewpoint of measurement theory, which is broadly concerned with the problem of numerically measuring a qualitative relation. Suppose there is a numeric function $f(x)$ that assigns real numbers to the objects in a lattice. These numbers represent the measurements. An ordinal measurement of the lattice order $\succcurlyeq$ requires that $x \succ y$ implies $f(x) \succ f(y)$. This type of measurement is only one-way. A two-way measurement requires that $x \succ y$ if and only if $f(x) \succ f(y)$. We can perform two-way measurements only upon total orders. In as much as the lattice order is partial, we cannot usually achieve a two-way measurement. We also want to extract information about the logical operators from the one-way measurement of logical order. This task is feasible only to the extent that we can reduce the algebraic structure into ordinal qualities.

When we evoke a numeric function to measure the lattice order, we refer to it as a *truth function*. Since the lattice is bounded, it is usually calibrated (without loss of generality) so that $f(\top) = 1$ and $f(\bot) = 0$. These bounds suggest that there is a degree-of-truth interpretation of the truth function, but a strictly ordinal approach does not provide a meaningful "degree of truth" for an interval-ratio scale of measurement. The operations of disjunction and conjunction are strictly ordinal. The involution is partly ordinal by virtue of the law of polarity. Only the law of double negation contains any trace of something nonordinal.

3.1 Canonical score

We next describe a method of constructing a truth function that works for any finite lattice. For any fixed object $a \in X$, define the *principal down-set* $D(a)$ by

$$D(a) := \{x \in X : a \succcurlyeq x\},$$

and define the *principal up-set* $U(a)$ by

$$U(a) := \{x \in X : x \succcurlyeq a\}.$$

Let $d(a)$ denote the frequency (or count) of objects in $D(a)$, and let $u(a)$ denote the frequency of objects in $U(a)$. Then the *canonical score* is defined by

$$\Phi(x) := \frac{d(x) - 1}{u(x) + d(x) - 2}.$$

The special numbers, 1 and 2, that appear in this formula are merely scaling constants. We can also think of it in terms of the relative weights of $d(x)$ and $u(x)$. The calibrated weights $d(x)-1$ and $u(x)-1$ count the number of objects strictly below and strictly above the object x.

https://doi.org/10.1515/9783112235812-003

Theorem 3.1. *Suppose X is a finite lattice. Then the canonical score* Φ *is a one-way measurement of the lattice order such that*
(1) $\Phi(\top) = 1$ *and* $\Phi(\bot) = 0$;
(2) $\Phi(x \vee y) \geq \max\{\Phi(x), \Phi(y)\}$;
(3) $\Phi(x \wedge y) \leq \min\{\Phi(x), \Phi(y)\}$.

The inequalities in conditions (2) and (3) are equalities if and only if x and y are comparable.

We can universally replace the inequalities in (2) and (3) with equalities only in the special case when the lattice order is total. In this case, the canonical score is a monomorphism. The canonical score also handles the involution elegantly.

Lemma 3.2. *If X is a finite de Morgan algebra, then* $\Phi({\sim} x) = 1 - \Phi(x)$.

By definition, a *canonical median* is an object x such that $\Phi(x) = 1/2$. Notice that x is a canonical median if and only if ${\sim} x$ is a canonical median. It is also clear that a self-negated object is a canonical median. The next result suggests that the canonical median is located in the "middle" of the algebra.

Lemma 3.3. *Suppose X is a finite lattice. Then an object x is a canonical median if and only if* $u(x) = d(x)$.

The canonical median does not necessarily exist. For example, a discrete total order has a median if and only if it contains an odd number of objects. However, it does provide a crude sense of valence in the algebra. We thus associate objects x such that $\Phi(x) > 1/2$ with "positive valence" and those objects such that $\Phi(x) < 1/2$ with "negative valence." However, this notion of valence is not fully satisfactory unless we are working with a total order.

The next computational result is sometimes useful.

Lemma 3.4. *Suppose X is a finite lattice. Then* $\Phi(x) = \Phi(y)$ *if and only if*

$$\frac{d(x) - 1}{u(x) - 1} = \frac{d(y) - 1}{u(y) - 1}.$$

It is noteworthy that $d(x) - 1$ counts the number of objects strictly below x; and that $u(x) - 1$ counts the number of objects strictly above x.

3.2 Standard score

There is a simpler way of measuring the order of a modular de Morgan algebra. In general, a (discrete) lattice satisfies the *Jordan–Dedekind chain condition* provided that all maximal chains between two endpoints have the same length (i. e., the same number of objects in the chain). A finite modular lattice always satisfies this condition. Furthermore, for every $x \in X$, there exists a maximal chain from bottom $\bot$ to top $\top$ that contains

x. We can therefore measure $x \in X$ unambiguously by its rank along one of these chains. Let $r(x)$ denote the rank of the object $x \in X$. For example, suppose

$$\top = x_m \succ x_{m-1} \succ \cdots \succ x_1 = \bot$$

is a maximal chain of length m. Then the rank $r(x_i) = i$ is computed by counting the number of objects in the chain from $\bot$ to x_i. The rank function in a finite modular lattice also satisfies the following property:

$$r(x \vee y) + r(x \wedge y) = r(x) + r(y).$$

We refer to this type of relation as the *diamond equation*. Moreover, a finite lattice is modular if and only if it satisfies the Jordan–Dedekind condition and the rank function satisfies the diamond equation [7].

Example. Figure 3.1 is a lattice X that satisfies the Jordan–Dedekind condition.

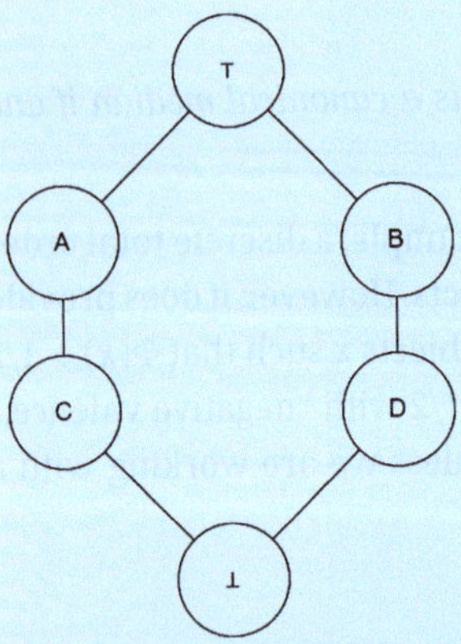

Figure 3.1: A nonmodular lattice with maximal chains of equal length.

Notice that $r(\top) + r(\bot) \neq r(A) + r(B)$. It thus violates the "diamond-like property" of the rank function for a modular lattice.

We next rescale the ranks. Suppose the (finite) lattice is modular and every maximal chain has length m. Then the *standard score* of $x \in X$ is defined by

$$Z(x) := \frac{r(x) - 1}{m - 1}.$$

The standard score has various properties like the canonical score.

Theorem 3.5. *Suppose X is a finite modular lattice. Then the standard score Z is a one-way measurement of the lattice order such that*

(1) $Z(\top) = 1$ *and* $Z(\bot) = 0$;
(2) $Z(x \vee y) \geq \max\{Z(x), Z(y)\}$;
(3) $Z(x \wedge y) \leq \min\{Z(x), Z(y)\}$;

The inequalities in conditions (2) and (3) are equalities if and only if x and y are comparable.

This result reads like Theorem 3.1 with Z replacing Φ. In parallel with Lemma 3.2, the standard score also handles involution nicely.

Lemma 3.6. *If X is a finite modular de Morgan algebra, then* $Z({\sim}\, x) = 1 - Z(x)$.

The above results about the standard score ultimately rely only on the Jordan–Dedekind chain condition. The standard score in a finite modular lattice also satisfies the *diamond equation*:

$$Z(x \vee y) + Z(x \wedge y) = Z(x) + Z(y).$$

We next redefine a median as the median of a maximal chain. Thus, the object x is a *standard median* provided that it solves $r(x) = (m + 1)/2$. It follows that the standard score of a standard median is $Z(x) = 1/2$.

Example. Consider the de Morgan algebra displayed in Figure 3.2.

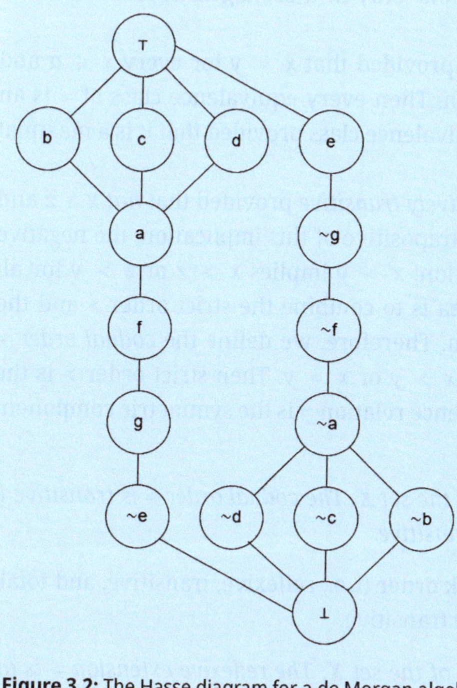

Figure 3.2: The Hasse diagram for a de Morgan algebra with multiple chains.

It is easy to verify that $\{a, \sim a\}$ is an opposite pair of canonical medians, but they are not standard medians. On the other hand, the pair $\{f, \sim f\}$ is an opposite pair of standard medians, but they are not canonical medians.

There is an important type of object that is both a canonical and standard median.

Lemma 3.7. *Suppose the de Morgan algebra X is finite. If the object $x \in X$ is self-negating, i. e., $\sim x = x$, then x is a canonical median. Furthermore, if X is also modular, then x is a standard median.*

Hence, we sometimes refer to a self-negating object as a "self-negating median." Medians need not be self-negating. Notice that if object x is one type of median, then $\sim x$ is a median of the same type. Another possibility is a pair of medians, x and $\sim x$, such that $x \parallel \sim x$ (see the above example).

3.3 Negative transitivity

The reflexive extension of the incomparability relation $\parallel$ is defined by $x \simeq y$ if and only if $x \parallel y$ or $x = y$. This relation is reflexive and symmetric, but typically intransitive. Thus, it is not necessarily an equivalence relation. In this section, we study the conditions that guarantee its transitivity. In general, we refer to $\simeq$ as a *preequivalence relation*. We are entitled to call it an "equivalence relation" only in those algebras for which it is transitive.

We say that a subset α is an *antichain* provided that $x \simeq y$ for every $x \in \alpha$ and $y \in \alpha$. Suppose $\simeq$ is an equivalence relation. Then every equivalence class of $\simeq$ is an antichain. Conversely, the subset α is an equivalence class provided that it is a maximal antichain.

By definition, the strict order $\succ$ is *negatively transitive* provided that *not* $x \succ z$ and *not* $z \succ y$ implies *not* $x \succ y$. Taking the contrapositive of this implication, the negative transitivity of $\succ$ is equivalent to the condition: $x \succ y$ implies $x \succ z$ or $z \succ y$ for all $z \in X$. Another way to understand this idea is to combine the strict order $\succ$ and the preequivalence relation $\simeq$ into one relation. Therefore, we define the *codual order* $\succcurlyeq$ by the condition that $x \succcurlyeq y$ if and only if $x \succ y$ or $x \simeq y$. Then strict order $\succ$ is the asymmetric component and the preequivalence relation $\simeq$ is the symmetric component of $\succcurlyeq$.

Lemma 3.8. *Suppose $\succcurlyeq$ is a partial order of the set X. The codual order $\succcurlyeq$ is transitive if and only if the strict order $\succ$ is negatively transitive.*

Therefore, the codual order $\succcurlyeq$ is a weak order (i. e., reflexive, transitive, and total) if and only if the strict order $\succ$ is negatively transitive.

Theorem 3.9. *Suppose $\succcurlyeq$ is a partial order of the set X. The reflexive extension $\simeq$ is an equivalence relation if and only if the strict order $\succ$ is negatively transitive.*

We will use this concept to strengthen the measurement of the lattice order with the canonical score. Two-way measurement refers to the measurement of the codual order: $x \succcurlyeq y$ if and only if $\Phi(x) \geq \Phi(y)$.

Lemma 3.10. *Suppose the de Morgan algebra X is finite. If $\Phi(x) = \Phi(y)$, then $x \simeq y$.*

A parallel result holds for the standard score in a modular algebra, but we do not need it right now. We seek a two-way representation of the ordinal structure such that
(1) $x \succ y$ implies $\Phi(x) > \Phi(y)$;
(2) $x \simeq y$ implies $\Phi(x) = \Phi(y)$.

The first part of this scheme has already been established. Notice that the second part is the converse to Lemma 3.8. There are only three possibilities for a pair of objects x and y: (i) $x \succ y$; (ii) $y \succ x$; and (iii) $x \simeq y$. The one-way implications in this scheme are thus two-way:
(1) $x \succ y$ if and only if $\Phi(x) > \Phi(y)$;
(2) $x \simeq y$ if and only if $\Phi(x) = \Phi(y)$.

Theorem 3.11. *Suppose X is a finite lattice. The canonical score Φ two-way represents the lattice order if and only if the strict order $\succ$ is negatively transitive.*

We next study some ramifications of negative transitivity.

Lemma 3.12. *Suppose X is a lattice. If the strict order $\succ$ is negatively transitive, then X is modular.*

Consequently, the standard score is well defined in this case. Hence, we aim to devise a two-way representation scheme for the standard score as well. The next result is analogous to Lemma 3.10.

Lemma 3.13. *Suppose X is a finite modular de Morgan algebra. If $Z(x) = Z(y)$, then $x \simeq y$.*

Likewise, the next result is the analogue of Theorem 3.11.

Theorem 3.14. *Suppose X is a finite modular lattice. The standard score Z two-way represents the lattice order if and only if the strict order $\succ$ is negatively transitive.*

We have thus established that both the canonical score and standard score two-way measure the lattice order by assuming negative transitivity.

The next result examines the fine structure of a lattice satisfying negative transitivity.

Lemma 3.15. *Suppose X is a lattice such that strict order $\succ$ is negatively transitive. If $a \parallel b$, then $x \simeq a$ if and only if $a \wedge b \prec x \prec a \vee b$.*

This result sheds light upon the equivalence classes of the equivalence relation $\simeq$. As long as the equivalence class of a includes a second object b distinct from a, then

$\{x \in X : x \simeq a\} = \{x \in X : a \wedge b \prec x \prec a \vee b\}$. Otherwise, the equivalence class for a is merely the singleton $\{a\}$. The next result considers the existence of three distinct objects in the same equivalence class.

Corollary. *Suppose X is a lattice such that strict order $\succ$ is negatively transitive. If $a \parallel b \parallel c \parallel a$, then*

$$a \vee b = a \vee c = b \vee c;$$
$$a \wedge b = a \wedge c = b \wedge c.$$

Notice that the relation $a \parallel b \parallel c$ implies $a \simeq c$, by virtue of the transitivity of the equivalence relation $\simeq$. Thus, we also need to assume $c \parallel a$ to assure that the equivalence class $\{x \in X : x \simeq a\}$ has at least three distinct objects.

A key point of this lemma and its corollary is that $[a \wedge b, a \vee b]$ is a sublattice containing $\{x \in X : x \simeq a\}$. The endpoints of this interval can be written in various ways depending upon the availability of other objects in the equivalence class for a.

Example. The lattice in Figure 3.3 is negatively transitive with $\{x \in X : x \simeq a\} = \{a, b, c\}$.

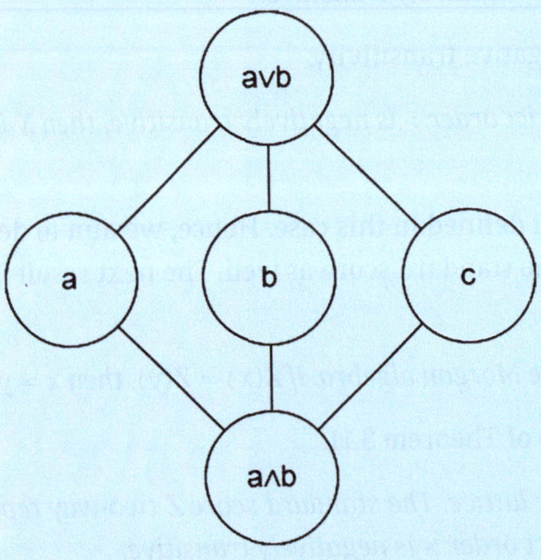

Figure 3.3: The Hasse diagram of a pentalemma.

This type of lattice underlies the "pentalemma." It is the simplest example of a nondistributive lattice. The pentalemma further assumes that two medians are opposites, and the third median is self-negated.

In general, a negatively transitive lattice consists of sublattices similar to a pentalemma with many medians, but stacked vertically. The next lemma provides the basic reason.

Lemma 3.16. *Suppose X is a lattice such that strict order $\succ$ is negatively transitive. If $a \parallel b$, then $\{x \in X : x \simeq a \vee b\}$ and $\{x \in X : x \simeq a \wedge b\}$ are singletons.*

According to the Birkhoff criterion, a lattice is distributive if and only it does not contain a sublattice isomorphic to the lattice of the pentalemma. Therefore, a negatively transitive lattice is distributive if and only if each equivalence class $\langle a \rangle$ contains at most two objects. This narrows the possibilities considerably, but it does include the diamond lattice and the total order. The general pattern in a Hasse diagram consists of vertically stacked diamond lattices connected by solitary vertical chains.

3.4 Proofs to this chapter

Theorem 3.1. *Suppose X is a finite lattice. Then the canonical score Φ is a one-way measurement of the lattice order such that*
(1) $\Phi(\top) = 1$ *and* $\Phi(\bot) = 0$;
(2) $\Phi(x \vee y) \geq \max\{\Phi(x), \Phi(y)\}$;
(3) $\Phi(x \wedge y) \leq \min\{\Phi(x), \Phi(y)\}$.

The inequalities in conditions (2) and (3) are equalities if and only if x and y are comparable.

Proof. First, $x \succ y$ implies $d(x) > d(y)$ and $u(y) > u(x)$. Let $a = d(x) - 1$, $b = d(y) - 1$, $c = u(y) - 1$, and $d = u(x) - 1$. It follows by algebra of arithmetic that $\frac{a}{d+a} > \frac{b}{c+b}$. Thus, we obtain $\Phi(x) > \Phi(y)$. Therefore, we conclude that Φ is a one-way measurement of the lattice order. We next prove its three basic lattice properties:

(1)

$$u(\top) = 1 \text{ implies } \Phi(\top) = \frac{d(\top) - 1}{u(\top) + d(\top) - 2} = 1, \quad \text{and}$$

$$d(\bot) = 1 \text{ implies } \Phi(\bot) = \frac{0}{u(\bot) - 1} = 0.$$

(2) Since $x \vee y \succcurlyeq x$, we obtain $\Phi(x \vee y) \geq \Phi(x)$; and since $x \vee y \succcurlyeq y$, we also obtain $\Phi(x \vee y) \geq \Phi(y)$. Thus, we obtain $\Phi(x \vee y) \geq \max\{\Phi(x), \Phi(y)\}$. We next establish the condition for equality. First, assume x and y are comparable. Without loss of generality, we assume $x \succcurlyeq y$. It follows that $\Phi(x) \geq \Phi(y)$ and $x = x \vee y$. Thus, condition (2) reduces to an equality. On the other hand, assume x and y are incomparable. Then $x \vee y \succ x$ and $x \vee y \succ y$ implies $\Phi(x \vee y) > \Phi(x)$ and $\Phi(x \vee y) > \Phi(y)$, respectively. Therefore, $\Phi(x \vee y) > \max\{\Phi(x), \Phi(y)\}$.

(3) The proof of condition (3), including the case of equality, is similar to the proof of condition (2). □

Lemma 3.2. *If X is a finite de Morgan algebra, then* $\Phi({\sim} x) = 1 - \Phi(x)$.

Proof. Since $x \succcurlyeq y$ if and only if ${\sim} y \succcurlyeq {\sim} x$ by virtue of the law of polarity, we obtain $d({\sim} x) = u(x)$ and $u({\sim} x) = d(x)$. It immediately follows by algebra of arithmetic that $\Phi({\sim} x) = \frac{u(x)-1}{u(x)+d(x)-2} = 1 - \Phi(x)$. □

Lemma 3.3. *Suppose X is a finite lattice. Then an object x is a canonical median if and only if $u(x) = d(x)$.*

Proof. This result is a simple exercise in arithmetic. It is easy to verify that $\frac{u(x)-1}{u(x)+d(x)-2} = \frac{1}{2}$ if and only if $u(x) = d(x)$ by cross multiplication. □

Lemma 3.4. *Suppose X is a finite lattice. Then $\Phi(x) = \Phi(y)$ if and only if*

$$\frac{d(x)-1}{u(x)-1} = \frac{d(y)-1}{u(y)-1}.$$

Proof. This result is another simple exercise in arithmetic. The equality $\Phi(x) = \Phi(y)$ expands into the equation

$$\frac{d(x)-1}{u(x)+d(x)-2} = \frac{d(y)-1}{u(y)+d(y)-2}.$$

One only needs to cross multiply and rearrange the resultant expression. This task is simplified by setting $a(x) = d(x) - 1$ and $b(x) = u(x) - 1$. □

Theorem 3.5. *Suppose X is a finite modular lattice. Then the standard score Z is a one-way measurement of the lattice order such that*
(1) $Z(\top) = 1$ *and* $Z(\bot) = 0$;
(2) $Z(x \vee y) \geq \max\{Z(x), Z(y)\}$;
(3) $Z(x \wedge y) \leq \min\{Z(x), Z(y)\}$.

The inequalities in conditions (2) and (3) are equalities if and only if x and y are comparable.

Proof. Suppose $x \succ y$. Construct a maximal chain from $\bot$ to $\top$ in three pieces: a maximal chain from $\bot$ to y; a maximal chain from y to x; and a maximal chain from x to $\top$. Following this chain, it is clear that x has a higher rank than y. Thus, we obtain $Z(x) > Z(y)$.

(1) The top is ranked m and the bottom is ranked 1. A direct computation yields $Z(\top) = 1$ and $Z(\bot) = 0$.

(2) Since $x \vee y \succcurlyeq x$ and $x \vee y \succcurlyeq y$, one-way measurement yields the inequalities $Z(x \vee y) \geq Z(x)$ and $Z(x \vee y) \geq Z(y)$. We thus obtain the inequality $Z(x \vee y) \geq \max\{Z(x), Z(y)\}$. This inequality reduces to equality if and only if $x \vee y = x$ or $x \vee y = y$. In the former case, we must have $x \succcurlyeq y$, and in the latter case, we must have $y \succcurlyeq x$.

(3) This inequality is the dual to inequality (2). □

Lemma 3.6. *If X is a finite modular de Morgan algebra, then $Z(\sim x) = 1 - Z(x)$.*

Proof. Let m denote the length of a maximal chain. If x has rank i, then a maximal chain from x to $\top$ has length $m - i + 1$. Applying the polar law, the involution of this maximal chain from x to $\top$ is a maximal chain from $\bot$ to $\sim x$. Thus, $\sim x$ has rank $m-i+1$. Therefore, the sum of the ranks of x and $\sim x$ is $m+1$. Since the standard score is a linear function of rank, we directly compute

$$Z(x) + Z(\sim x) = \frac{i-1}{m-1} + \frac{(m-i+1)-1}{m-1} = 1.$$

The formula is derived by rearranging the terms. □

Lemma 3.7. *Suppose the de Morgan algebra X is finite. If the object $x \in X$ is self-negating, i. e., $\sim x = x$, then x is a canonical median. Furthermore, if X is also modular, then x is a standard median.*

Proof. The fact that a self-negated object is a canonical median follows from Lemma 5.2. Likewise, it is a standard median in a modular algebra by virtue of Lemma 3.6. □

Lemma 3.8. *Suppose $\succcurlyeq$ is a partial order of the set X. The codual order $\succcurlyeq$ is transitive if and only if the strict order $\succ$ is negatively transitive.*

Proof. Notice that $x \succcurlyeq y$ if and only if not $y \succ x$. The transitivity of $\succcurlyeq$ is defined by the condition that $x \succcurlyeq y$ and $y \succcurlyeq z$ implies $x \succcurlyeq z$. This definition is equivalent to the condition that not $y \succ x$ and not $z \succ y$ implies not $z \succ x$. This condition is the negative transitivity of $\succ$. □

Theorem 3.9. *Suppose $\succcurlyeq$ is a partial order of the set X. The reflexive extension $\simeq$ is an equivalence relation if and only if the strict order $\succ$ is negatively transitive.*

Proof. Suppose $\simeq$ is an equivalence relation. Claim: $\simeq$ is transitive implies $\succcurlyeq$ is transitive, i. e., $x \succcurlyeq y$ and $y \succcurlyeq z$ implies $x \succcurlyeq z$. We need to analyze four distinct possibilities: (1) Assume $x \succ y$ and $y \succ z$. Then we obtain $x \succ z$ by virtue of the transitivity of $\succ$. (2) Assume $x \simeq y$ and $y \simeq z$. Then we obtain $x \simeq z$ by virtue of the transitivity of $\simeq$. (3) Assume $x \succ y$ and $y \simeq z$. Then $z \succ x$ is impossible, because it implies $z \succ y$; and $z \simeq x$ is impossible, because it implies $y \simeq x$. Therefore, we obtain $x \succ z$. (4) Assume $x \simeq y$ and $y \succ z$. Then $z \succ x$ is impossible, because it implies $y \succ x$; and $z \simeq x$ is impossible, because it implies $z \simeq y$. Therefore, we obtain $x \succ z$. Since we have verified transitivity in all four possible scenarios, we have verified the claim that $\simeq$ is transitive implies $\succcurlyeq$ is transitive.

Conversely, assume that $\succ$ is negatively transitive. Then it follows from Lemma 3.8 that the codual order $\succcurlyeq$ is transitive. Thus, its symmetric component $\simeq$ is transitive. Since $\simeq$ is also symmetric and reflexive, it is an equivalence relation. □

Lemma 3.10. *Suppose the de Morgan algebra X is finite. If $\Phi(x) = \Phi(y)$, then $x \simeq y$.*

Proof. Since Φ one-way represents $\succcurlyeq$, we know that $x \succ y$ implies $\Phi(x) > \Phi(y)$ and $y \succ x$ implies $\Phi(y) > \Phi(x)$. Therefore, $\Phi(x) = \Phi(y)$ implies $x \simeq y$. □

Theorem 3.11. *Suppose X is a finite lattice. The canonical score Φ two-way represents the lattice order if and only if the strict order $\succ$ is negatively transitive.*

Proof. Assume that Φ two-way represents $\succcurlyeq$. Then $a \simeq b$ implies $\Phi(a) = \Phi(b)$. But $\Phi(a) = \Phi(b)$ directly implies $a \nsucc b$ and $b \nsucc a$, so that $a \simeq b$. Since $a \simeq b$ if and only if $\Phi(a) = \Phi(b)$, it follows that $\simeq$ is an equivalence relation. Therefore, $\succ$ is negatively transitive by virtue of Lemma 3.8.

Conversely, assume that $\succ$ is negatively transitive. We claim that $a \simeq b$ implies

$$\{x : x \succ a\} = \{x : x \succ b\} \quad \text{and} \quad \{x : x \prec a\} = \{x : x \prec b\}.$$

Since this result is trivial when $a = b$, we only need to check it when $a \parallel b$. Since $\succ$ is negatively transitive, $x \succ a$ implies $b \succ a$ or $x \succ b$. Thus, $x \succ a$ implies $x \succ b$. By symmetry, $x \succ b$ implies $x \succ a$. The second part follows by duality.

It immediately follows that $a \simeq b$ implies $u(a) = u(b)$ and $d(a) = d(b)$. Therefore, $a \simeq b$ implies $\Phi(a) = \Phi(b)$. We also know from Theorem 3.1 that $x \succ y$ implies $\Phi(x) > \Phi(y)$. Therefore, Φ two-way represents $\succcurlyeq$. ☐

Lemma 3.12. *Suppose X is a lattice. If the strict order $\succ$ is negatively transitive, then X is modular.*

Proof. Suppose that the lattice order $\succ$ is negatively transitive, i. e., $x \prec y$ implies $x \prec z$ or $z \prec y$. We must show that $x \preccurlyeq y$ implies $x \vee (y \wedge z) = y \wedge (x \vee z)$. The result holds when $x = y$ by virtue of the absorption laws. We thus only need to verify modularity in the case when $x \prec y$. There are two cases to consider. In the first, assume that $x \prec y$ and $x \prec z$. Since $x \preccurlyeq y \wedge z$, it follows that $x \vee (y \wedge z) = y \wedge z$. Since $x \prec z$, it also follows that $y \wedge (x \vee z) = y \wedge z$. This verifies the first case. In the second, assume that $x \prec y$ and $z \prec y$. Since $z \prec y$, it follows that $x \vee (y \wedge z) = x \vee z$. Since $x \vee z \preccurlyeq y$, it also follows that $y \wedge (x \vee z) = x \vee z$. This verifies the second case. ☐

Lemma 3.13. *Suppose X is a finite modular de Morgan algebra. If $Z(x) = Z(y)$, then $x \simeq y$.*

Proof. Assume $Z(x) = Z(y)$. Notice that $x \succ y$ implies $Z(x) > Z(y)$ and that $y \succ x$ implies $Z(y) > Z(x)$. Therefore, we obtain $x \nsucc y$ and $y \nsucc x$, i. e., $x \simeq y$. ☐

Theorem 3.14. *Suppose X is a finite modular lattice. The standard score Z two-way represents the lattice order if and only if the strict order $\succ$ is negatively transitive.*

Proof. First, assume the strict order $\succ$ is negatively transitive. Since Z one-way measures $\succcurlyeq$ by virtue of Theorem 3.5, we have already confirmed that $x \succ y$ implies $Z(x) > Z(y)$. Assume $Z(x) > Z(y)$. Applying Lemma 3.12 to the finite lattice, every maximal chain from y to the top $\top$ has the same length. Therefore, there exists an element $w \in X$ such that $w \succ y$ and w has the same rank as x. Thus, $Z(x) = Z(w)$, which implies $x \simeq w$ by virtue of Lemma 3.13. We know from Lemma 3.8 that the codual order is transitive. Thus, $x \simeq w \succ y$ implies $x \succ y$. In summary, $Z(x) > Z(y)$ implies $x \succ y$. Therefore, $x \simeq y$ if and only if $Z(x) = Z(y)$.

The converse is straightforward. Assume that Z two-way measures $\succcurlyeq$. Assume $x \succcurlyeq y \succcurlyeq w$. It follows that $Z(x) \geq Z(y) \geq Z(w)$. Thus, $Z(x) \geq Z(w)$ implies $x \succcurlyeq w$. This verifies that the codual order $\succcurlyeq$ is transitive. The negative transitivity of the strict order $\succ$ follows from Lemma 3.8. □

Lemma 3.15. *Suppose X is a lattice such that strict order $\succ$ is negatively transitive. If $a \parallel b$, then $x \simeq a$ if and only if $a \wedge b \prec x \prec a \vee b$.*

Proof. Suppose $\succ$ is negatively transitive and $a \parallel b$ throughout. First, we assume $a \wedge b \prec x \prec a \vee b$. We will show that $x \simeq a$. Contrary to hypothesis, assume $x \not\simeq a$, so that $x \succ a$ or $a \succ x$. In the former case, we obtain the triplet $x \succ a \parallel b$. Evoking the negative transitivity of $\succ$, we deduce $x \succ b$. Since $x \succ a$ and $x \succ b$, we find $x \succcurlyeq a \vee b$, which contradicts the assumption that $x \prec a \vee b$. In the latter case, we obtain the triplet $b \parallel a \succ x$. Evoking the negative transitivity of $\succ$, we deduce $b \succ x$. Since $a \succ x$ and $b \succ x$, we find $a \wedge b \succcurlyeq x$, which contradicts the assumption $a \wedge b \prec x$. Since both contrary cases lead to a contradiction, we conclude that $x \simeq a$.

Conversely, assume $x \simeq a$. We will show that $a \wedge b \prec x \prec a \vee b$. Consider the endpoint $a \vee b$. Contrary to hypothesis, assume $x \succcurlyeq a \vee b$. Evoking the transitivity of $\succcurlyeq$, $a \simeq x \succcurlyeq a \vee b$ implies $a \succcurlyeq a \vee b$, which contradicts $a \prec a \vee b$. Therefore, we obtain $x \prec a \vee b$, as we wished to show. Consider the other endpoint $a \wedge b$. Contrary to hypothesis, assume $x \preccurlyeq a \wedge b$. Evoking the transitivity of $\succcurlyeq$, $a \simeq x \preccurlyeq a \wedge b$ implies $a \preccurlyeq a \wedge b$, which contradicts $a \wedge b \prec a$. Therefore, we obtain $a \wedge b \prec x$, as we wished to show. In summary, we have verified $a \wedge b \prec x \prec a \vee b$. □

Corollary. *Suppose X is a lattice such that strict order $\succ$ is negatively transitive. If $a \parallel b \parallel c \parallel a$, then*

$$a \vee b = a \vee c = b \vee c;$$

$$a \wedge b = a \wedge c = b \wedge c.$$

Proof. Since $c \in \{x : x \simeq a\}$, we obtain $a \wedge b \prec c \prec a \vee b$. Thus,

$$(1) \quad a \wedge b \preccurlyeq a \wedge c \prec a \vee c \preccurlyeq a \vee b.$$

Since $b \in \{x : x \simeq c\}$, we also obtain $a \wedge c \prec b \prec a \vee c$. Thus,

$$(2) \quad a \wedge c \preccurlyeq a \wedge b \prec a \vee b \preccurlyeq a \vee c.$$

Comparing (1) and (2), we obtain $a \wedge b = a \wedge c$ and $a \vee b = a \vee c$. □

Lemma 3.16. *Suppose X is a lattice such that strict order $\succ$ is negatively transitive. If $a \parallel b$, then $\{x \in X : x \simeq a \vee b\}$ and $\{x \in X : x \simeq a \wedge b\}$ are singletons.*

Proof. Suppose $a \parallel b$. Select $c \simeq a \vee b$. Since $a \vee b \succ a$ and $a \vee b \succ b$, we infer that $c \succ a$ and $c \succ b$ from negative transitivity. Thus, $c \succcurlyeq a \vee b$ and $c \simeq a \vee b$ imply $c = a \vee b$. Therefore, we obtain $\{x \in X : x \simeq a \vee b\}$ is a singleton. The proof that $\{x \in X : x \simeq a \wedge b\}$ is a singleton is similar. □

3.5 Exercises

(1) Let $[0,1]$ denote the unit interval. Define two binary operators by $x \vee y := \max\{x,y\}$ and $x \wedge y := \min\{x,y\}$. Verify that $[0,1]$ is a bounded lattice. Define the unary operator $\sim x := 1-x$. Verify that this operator an involution, and thus the unit interval $[0,1]$ is a de Morgan algebra. (This is called fuzzy logic.)

(2) Suppose X is a finite de Morgan algebra and $f : X \to [0,1]$ is a map such that $f(\sim x) = 1 - f(x)$. Demonstrate that

$$f(x \vee y) = \max\{f(x), f(y)\}$$

if and only if

$$f(x \wedge y) = \min\{f(x), f(y)\}.$$

(3) Compute the canonical and standard scores of all the elements in the pentagonal lattice.

(4) Suppose X is a finite de Morgan algebra. Demonstrate that the following two conditions are mutually equivalent:
  (i) $a \simeq \sim a$ implies a is a canonical median, i. e., $\Phi(a) = 1/2$;
  (ii) $\Phi(a) > 1/2$ implies $a \succ \sim a$.

(5) Suppose X is a finite de Morgan algebra. Define a new measure of the lattice order by the formula $\Psi(x) := d(x)/u(x)$. (i) Demonstrate that Ψ one-way measures the lattice order $\succcurlyeq$. (ii) Demonstrate that $\Psi(\sim x) = 1/\Psi(x)$. (iii) Demonstrate that x is a canonical median if and only if $\Psi(x) = 1$.

(6) Suppose X is a finite modular lattice. If x and y are standard medians, verify that $Z(x \vee y) = 1 - Z(x \wedge y)$.

(7) Suppose X is a finite modular lattice. Verify the following identity:

$$Z(x \vee y) - \max\{Z(x), Z(y)\} = \min\{Z(x), Z(y)\} - Z(x \wedge y).$$

(8) Suppose X is a de Morgan algebra. Verify that $x \succcurlyeq y$ if and only if $\sim y \succcurlyeq \sim x$, and that $x \succcurlyeq y$ if and only if $y \nsucc x$.

(9) Suppose X is a distributive lattice. Demonstrate that $\succ$ is negatively transitive if and only if $x \parallel z$ and $y \parallel z$ imply $x = y$.

(10) Suppose X is a de Morgan algebra such that the strict order $\succ$ is negatively transitive. If $a \parallel b$, demonstrate that $x \simeq a$ if and only if $a \wedge b \prec x \prec a \vee b$.

4 The tetralemma

A tetralemma consists of four premises: it exists, it does not exist, both, and neither. These were regarded as the only four possibilities among Buddhist logicians. A common method of analysis pioneered by Nagarjuna was to reject the four premises one-by-one, concluding that something is "empty." However, we think that this is a one-sided use of the tetralemma. In particular, T'ien-tai used the tetralemma to classify Buddhist teachings and organize basic concepts like the Four Noble Truths. Emptiness is only one facet of ultimate reality.

The opposite of emptiness is nonemptiness. We give equal weight to both of these ideas in the logical design of the tetralemma. Intuitively, we tend to associate emptiness with falsehood and nonemptiness with truth. Yet, both emptiness and nonemptiness are partial truths integrated into the Middle Way. Although we can locate the middle region of the tetralemma graphically, it is probably best to think of the Middle Way as encompassing all four viewpoints.

4.1 The model

By definition, a *truth-value* is any element belonging to a *truth lattice*, i. e., an instrument for assigning and evaluating "truth." The tetralemma can be visualized as a diamond lattice with four truth-values corresponding to the four premises: existence, nonexistence, both, and neither. However, the arrangement of the four premises into a lattice is enigmatic. We have already suggested that "both" is a disjunction and "neither" is a conjunction. We thus arrange the four premises into an algebraic configuration (see Figure 4.1) that looks like a Hasse diagram, but its nodes are variables.

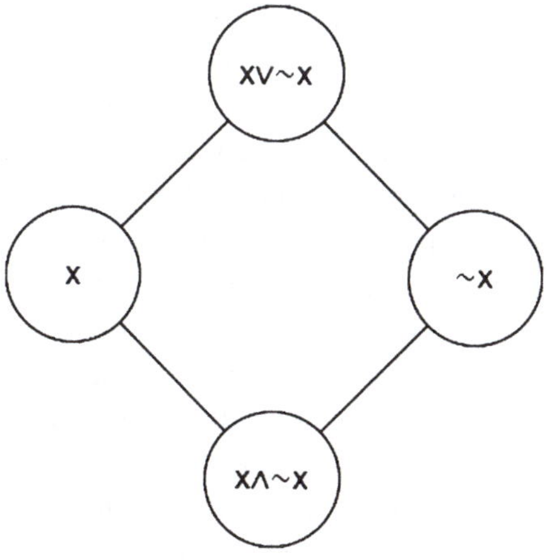

Figure 4.1: The algebraic tetralemma.

https://doi.org/10.1515/9783112235812-004

The algebraic tetralemma provides information about the involution that the diamond lattice alone does not. We begin with two incomparable opposites, x and $\sim x$, and take their disjunction as "both" and their conjunction as "neither." The negated disjunction $\sim(x \vee \sim x)$ is literally "neither existence nor nonexistence." In brief, it is "empty." Nagarjuna's identity (see Lemma 2.3) converts emptiness into the "oneness of existence and nonexistence," i. e., $x \wedge \sim x$. Thus, the top and bottom of the tetralemma are opposites.

In conventional logic, the conjunction $x \wedge \sim x$ is called a "contradiction," but this terminology is inadequate in the de Morgan algebra. We refer to it as the "oneness of x and $\sim x$" instead. The oneness of two opposites has long been a mainstay of mysticism, but we now see that the concept fits well into the logic of the de Morgan algebra. We speculate that Nagarjuna reasoned in the following way. First, he observed that all phenomena are empty (i. e., $\sim(x \vee \sim x)$). The reason that they are empty is because their existence depends upon the existence of everything else. This is the oneness of self and other (i. e., $x \wedge \sim x$). This identity between emptiness and dependent origination is what he called the Middle Way.

Hegel's dialectic logic also fits neatly into this scheme. This logic creates a new idea from the synthesis of two opposing ideas, the thesis and the antithesis. We simply interpret x as the *thesis*, $\sim x$ as the *antithesis*, and $x \wedge \sim x$ as the *synthesis*. In particular, Hegel identified "becoming" as the synthesis of "being" and "nothing." Alternatively, we say that "temporary existence" is the oneness of existence and nonexistence. Hegel's idea of the synthesis is therefore closely related to Nagarjuna's concept of emptiness.

Let us review the ontological meaning to the four premises. First, we associate x with "existence" and $\sim x$ with "nonexistence." In ordinary logic, one associates existence with "true" and nonexistence with "false," ordering truth over falsehood. The goal of this logic is to preserve truth while making inferences. The old way of thinking does not work here. We have set up x and $\sim x$ as two incomparable opposites. Neither one of them has more claim to truth than the other. Next, we associate $x \wedge \sim x$ with "emptiness," and we associate $x \vee \sim x$ with "nonemptiness." We must probe more deeply into the meaning of existence to grasp this distinction. Ancient philosophers frequently associated "existence" with a permanent irreducible substance. For example, the ancient theory of atoms was such a theory that endured until the twentieth century. In this context, emptiness means there is no permanent substance within phenomena. However, they still have a temporary existence. On the other hand, "nonemptiness" is not the same as existence. Nonemptiness recognizes the hidden potential with conventional existence. There is a synergy within $x \vee \sim x$ that cannot be explained by x and $\sim x$ separately.

In Mahayana Buddhism, the semantics are upgraded. The concepts of existence and nonexistence are discarded completely, replacing existence with nonemptiness and nonexistence with emptiness. Thus, the "tetralemma of emptiness" is constructed from a new version of the four premises: emptiness, nonemptiness, both emptiness and nonemptiness, and neither emptiness nor nonemptiness. The premise of "neither emptiness nor nonemptiness" is also called the "emptiness of emptiness," and the premise of

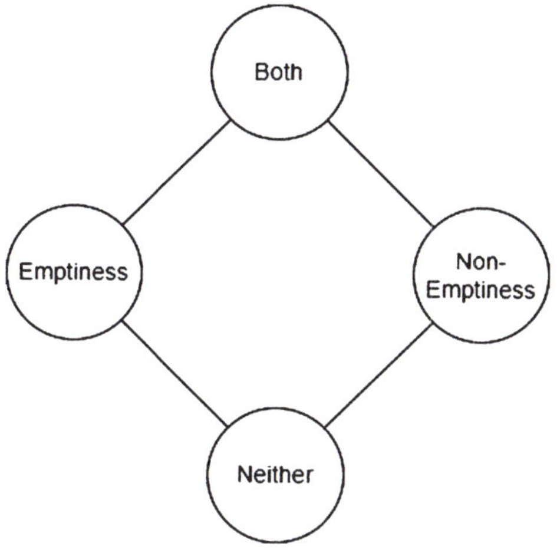

Figure 4.2: The tetralemma of emptiness.

both emptiness and nonemptiness is called the "nonemptiness of emptiness." The labels of the four nodes of the Hasse diagram are thus modified (see Figure 4.2).

Of course, the syntax of the tetralemma is the same. Therefore, we usually rely upon the "tetralemma of existence" for the basic ontological interpretation.

The notion of mathematically analyzing nondual logic may seem paradoxical to some readers. The whole idea is to reject the kind of logic that mathematics itself is based upon (i. e., classical logic). However, the methods that we use are based upon universal algebra. The syntax is algebraic even though the semantics are mystical. Thus, we do not appeal to mathematics merely as a metalanguage for logic; rather it is integrated in the formal design. Nonduality enters into this logic through the semantics. There is both formal semantics embodied into multiple truth-values and informal semantics interpreting what we are doing with nondual analogies.

4.2 Nominal analysis

There are only a few ways of organizing the four premises into a lattice. We will briefly discuss other possibilities under the proviso that each model must be a de Morgan algebra. This type of analysis is "nominal" in the sense that it does not depend upon the ontological interpretation of truth-value. However, informal semantics is the primary method for evaluating the different models. This is an exercise in the design of a logical system.

The simplest scheme stacks them in a vertical hierarchy (see Figure 4.3).

In this configuration, there are two pairs of opposites: existence versus nonexistence and emptiness versus nonemptiness. However, the syntax is different from the tetralemma, because both pairs are comparable. Of course, the disjunction of all four

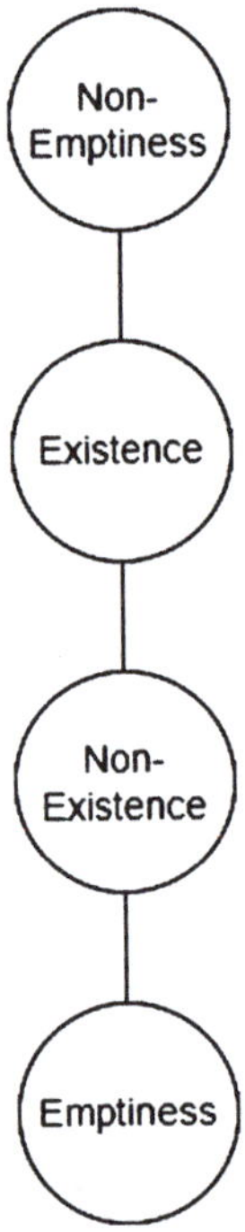

Figure 4.3: Crude model of the four premises.

truth-values is the top, and the conjunction of all four is the bottom. One can also exchange the relative positions of the truth-values in various ways as long as we respect the law of polarity. A basic question emerges: "Why are existence and nonexistence framed as incomparable in the logic of the tetralemma?" Perhaps the best reason is that it recovers the syntax of "both" and "neither" from the relation of existence and nonexistence. Thus, nonemptiness is the meaning assigned to "both" and emptiness is the meaning assigned to "neither." We cannot attain these distinctions with a total lattice order.

We next examine the four-valued logic devised by Nuel Belnap [8]. He associated the four truth-values with the power set of $\{T, F\}$ in the following manner:

Truth-Value	True	False	Both	Neither
Subset	$\{T\}$	$\{F\}$	$\{T,F\}$	$\emptyset$
Negation	$\{F\}$	$\{T\}$	$\{T,F\}$	$\emptyset$

Belnap explained truth-value as the internal state of a computer receiving input about the truth or falsity of an event from multiple information sources: The event is assigned the truth-value of $\{T\}$ provided that the data consistently reports it as true; it is assigned a truth-value of $\{F\}$ provided that the data consistently reports it as false; it is assigned a truth-value of $\{T, F\}$ provided that the data reports it as sometimes true and other times false; and it is assigned a truth-value of ∅ provided that there is no data reporting it either way. Naturally, the subset $\{T, F\}$ is interpreted as "both," and the empty subset ∅ is interpreted as "neither." We refer to this setup as the *Belnap representation scheme*. It leads to an interesting concept of negation, where one simply juxtaposes the symbols "T" and "F". Thus, $\{T\}$ and $\{F\}$ are opposites, but $\{T, F\}$ and ∅ are self-negated. In terms of the power set representation of the four truth values, this type of negation is element-wise instead of complementation.

He organized the four truth-values into a diamond lattice in two different ways. (1) The "Belnap information lattice" takes the truth-values of "true" and "false" as medians with a disjunction of "both" and a conjunction of "neither." This model is not a de Morgan algebra, simply because the top and bottom are not opposites. However, the lattice order is consistent with the logic of sets. (2) The "Belnap logical lattice" takes "both" and "neither" as medians with a disjunction of true and a conjunction of false. This model is a de Morgan algebra. However, the lattice order is not consistent with the logic of sets. In other words, the power set representation proposed by Belnap is a bit misleading, because it encourages one to think set-theoretically. Most importantly, the association of "both" with lattice disjunction and "neither" with conjunction seems lost.

We can also use the power set representation for the tetralemma. In this case, involution is set-theoretic complementation, disjunction is union, and conjunction is intersection. Complementation works with the set-theoretic representation of the tetralemma, because it is a Boolean algebra. Figure 4.4 is the new schematic of the Hasse diagram:

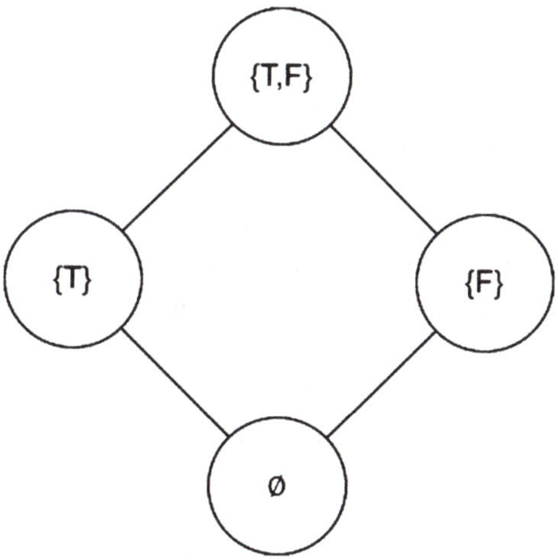

Figure 4.4: Power set representation of the tetralemma.

This setup can be derived from the Birkhoff's representation theorem [9], which represents a finite distributive lattice with the sublattice of a power set.

4.3 Four logic gates

There is another way to grasp the four premises. We now represent them as four logic gates: the gate of existence, the gate of nonexistence, the gate of emptiness, and the gate of nonemptiness. In algebraic terms, these gates are four unary operators: the *gate of existence* is the identity map $x \mapsto x$; the *gate of nonexistence* is the involution $x \mapsto \sim x$; the *gate of nonemptiness* is the map $x \mapsto x \vee \sim x$; and the *gate of emptiness* is the map $x \mapsto x \wedge \sim x$. These maps are typically hidden within a truth lattice by the colorful labels attached to the truth-values. The last two gates are transcendental. In a Boolean algebra, the gate of emptiness always leads to the bottom $\perp$, and the gate of nonemptiness always leads to the top $\top$. These gates behave differently in the more general de Morgan algebra.

We will focus upon the transcendental gates, because they are more difficult to see. The first problem is to determine which truth-values are in the codomain of these gates. By definition, the object a is *nonempty* provided that $a = x \vee \sim x$ for some $x \in X$; and the object a is *empty* provided that $a = x \wedge \sim x$ for some $x \in X$. These definitions are derived from the instrumental definitions of emptiness and nonemptiness, and they are strictly a matter of algebraic form. Regarding them as equations for the free variable x, they are easy to solve.

Theorem 4.1. *Suppose X is a de Morgan algebra and $a \in X$ is a fixed object. The equation $a = x \vee \sim x$ is solvable for some $x \in X$ if and only if $a \succcurlyeq \sim a$. Likewise, the equation $a = x \wedge \sim x$ is solvable for some $x \in X$ if and only if $\sim a \succcurlyeq a$.*

It follows from this theorem that empty and nonempty objects typically exist as pairs of comparable opposites, say $a \succcurlyeq \sim a$. In this case, object a is nonempty and object $\sim a$ is empty. However, this theorem further indicates that a self-negating object (i. e., $a = \sim a$) is *both* empty and nonempty (i. e., it can be represented as both forms). A fourth possibility is that an object is *neither* empty nor nonempty (i. e., it can be represented as neither of these forms).

Let us briefly examine empty and nonempty objects from the viewpoint of conventional logic. The bottom is an empty-only object. Any other empty object is conventionally thought to be a "contradiction," i. e., a violation of the law of noncontradiction. In this sense, the de Morgan algebra allows the possibility of some contradictions. On the other hand, the top is a nonempty-only. Any other nonempty object is thought to be an "intermediate existence" beyond existence and nonexistence, i. e., a violation of the law of excluded middle. Most importantly, an intermediate existence is the involution of a contradiction.

This analysis leads to a four-way classification of the objects in the general de Morgan algebra: empty-only, nonempty-only, both empty and nonempty, and neither empty nor nonempty. Clearly, an object a is *nonempty-only* provided that $a \succ \sim a$. Furthermore, a is *empty-only* if and only if $\sim a$ is nonempty-only. For example, the four truth-values in a tetralemma are easily classified this way: the top is nonempty-only; the bottom is empty-only; and the two opposite medians in the middle region are neither empty nor nonempty.

Lemma 4.2. *Suppose X is a de Morgan algebra. An object $a \in X$ is self-negated if and only if it is both empty and nonempty.*

We have already introduced the principal down-set $D(a)$ and the principal up-set $U(a)$. More generally, a *down-set* D is a subset of the lattice X such that $a \in D$ and $a \succcurlyeq x$ implies $x \in D$; and an up-set U is a subset of the lattice X such that $a \in U$ and $x \succcurlyeq a$ implies $x \in U$.

Theorem 4.3. *Suppose X is a de Morgan algebra. (1) The subset of all nonempty objects in X is an up-set, and the subset of all empty objects is a down-set. (2) Likewise, the subset of all nonempty-only objects in X is an up-set, and the subset of all empty-only objects is a down-set.*

This theory is also useful for uncovering subalgebras. The following result presents a basic method of finding subalgebras.

Lemma 4.4. *Suppose X is a de Morgan algebra. If $a \succ \sim a$ for some $a \in X$, then the interval $[\sim a, a]$ is a subalgebra of X.*

In turn, this interval of opposite endpoints can also contain various subalgebras. For example, the two endpoints $\{a, \sim a\}$ form a subalgebra isomorphic to the *Boolean truth lattice* (i. e., $1 \succ 0$) provided that $a \succ \sim a$. In contrast, any pair $\{a, \sim a\}$ of incomparable opposites is the basis of a subalgebra $\{a, \sim a, a \vee \sim a, a \wedge \sim a\}$ isomorphic to a tetralemma.

Lemma 4.5. *Suppose X is a Boolean algebra. An object $a \in X$ is nonempty if and only if $a = \top$, and it is empty if and only if $a = \bot$. All other objects in X (besides $\top$ and $\bot$) are neither empty nor nonempty.*

Consequently, all opposite pairs $\{a, \sim a\}$ in a Boolean algebra except $\{\top, \bot\}$ are incomparable, and constitute the basis of a subalgebra $\{a, \sim a, \top, \bot\}$ isomorphic to a tetralemma. There are no self-negating objects in a Boolean algebra.

Suppose $a \parallel \sim a$ are incomparable opposites in some de Morgan algebra X. The tetralemma $\{a, \sim a, a \vee \sim a, a \wedge \sim a\}$ itself is obviously a subalgebra of X.

We can also imagine the four nodes as the endpoints of four intervals: the interval $[\sim a, a \vee \sim a]$ is the opposite side of the interval $[a \wedge \sim a, a]$; and the interval $[a, a \vee \sim a]$ is the opposite side of the interval $[a \wedge \sim a, \sim a]$. This idea is illustrated in Figure 4.5. Moreover, the involution is a well-defined invertible map from the interval on one side to its opposite. Thus, it is a dual isomorphism between intervals on opposites of the tetralemma. This result is analogous to the diamond isomorphism theorem, but it does not rely upon the modular law.

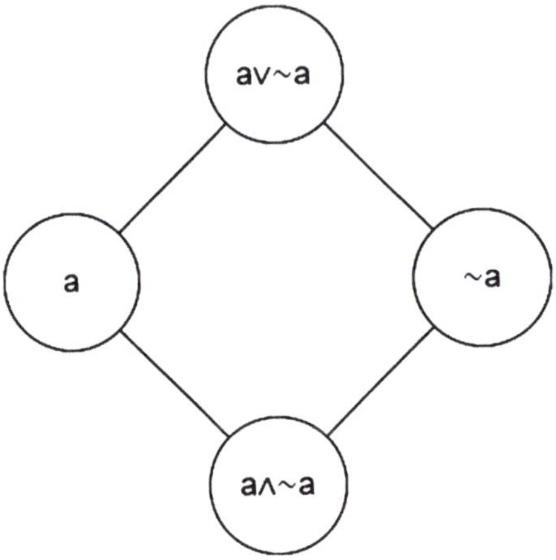

Figure 4.5: The tetralemma generated by incomparable opposites.

4.4 Proofs to this chapter

Theorem 4.1. *Suppose X is a de Morgan algebra and $a \in X$ is a fixed object. The equation $a = x \vee \sim x$ is solvable for some $x \in X$ if and only if $a \succcurlyeq \sim a$. Likewise, the equation $a = x \wedge \sim x$ is solvable for some $x \in X$ if and only if $\sim a \succcurlyeq a$.*

Proof. Assume $a \succcurlyeq \sim a$. Then we obtain $a = a \vee \sim a$, so that $x = a$ is the required form. Conversely, assume $a = x \vee \sim x$. There are two cases to consider. (1) If $x \parallel \sim x$, then we obtain $a = x \vee \sim x \succ x \wedge \sim x = \sim a$. (2) If $x \nparallel \sim x$, then without loss of generality we may assume $x \succcurlyeq \sim x$. Thus, $a = x \vee \sim x = x$, so that $a \succcurlyeq \sim a$ in this case. Notice that $a = x \vee \sim x$ if and only if $\sim a = a \wedge \sim a$. Thus, the solvability of the equation $a = x \wedge \sim x$ is equivalent to $a \succcurlyeq \sim a$. □

Lemma 4.2. *Suppose X is a de Morgan algebra. An object $a \in X$ is self-negated if and only if it is both empty and nonempty.*

Proof. Assume the object a is self-negated. Then $a = \sim a$ implies $a = a \wedge \sim a$ and $a = a \vee \sim a$. Thus, a is both empty and nonempty. Conversely, assume a is both empty and nonempty. Since a is nonempty, we obtain $a \succcurlyeq \sim a$. Since a is empty, we obtain $\sim a \succcurlyeq a$. Since the partial order $\succcurlyeq$ is antisymmetric, we conclude that $a = \sim a$. □

Theorem 4.3. *Suppose X is a de Morgan algebra. (1) The subset of all nonempty objects in X is an up-set, and the subset of all empty objects is a down-set. (2) Likewise, the subset of all nonempty-only objects in X is an up-set, and the subset of all empty-only objects is a down-set.*

Proof. (1) Let α denote the subset of all nonempty elements. Assume $a \in \alpha$ and $b \succcurlyeq a$. We obtain $a \succcurlyeq \sim a$ from Lemma 4.2, and $\sim a \succcurlyeq \sim b$ from the polarity law. Thus, we obtain

$$b \succcurlyeq a \succcurlyeq {\sim}a \succcurlyeq {\sim}b.$$

Since $b \succcurlyeq {\sim}b$, it follows from Lemma 4.2 that $b \in \alpha$. Therefore, α is an up-set. The element-wise involution of an up-set is a down-set. The conclusion is that the involuted subset ${\sim}\alpha := \{x \in X : {\sim}x \in \alpha\}$ of empty elements is a down-set.

(2) Let β denote the subset of all nonempty-only elements. Assume $a \in \beta$ and $b \succcurlyeq a$. Then $a \in \beta$ implies $a \succ {\sim}a$. In this case, we obtain

$$b \succcurlyeq a \succ {\sim}a \succcurlyeq {\sim}b.$$

Thus, $b \succ {\sim}b$, which implies $b \in \beta$. Therefore, β is an up-set and ${\sim}\beta$ is a down-set. □

Lemma 4.4. *Suppose X is a de Morgan algebra. If $a \succ {\sim}a$ for some $a \in X$, then the interval $[{\sim}a, a]$ is a subalgebra of X.*

Proof. Suppose $a \succ {\sim}a$. Select $x \in [{\sim}a, a]$ and $y \in [{\sim}a, a]$. Then $a \succcurlyeq x \succcurlyeq {\sim}a$ and $a \succcurlyeq y \succcurlyeq {\sim}a$ imply $a \succcurlyeq x \vee y \succcurlyeq x \wedge y \succcurlyeq {\sim}a$. Applying the polarity law, we find that $a \succcurlyeq x \succcurlyeq {\sim}a$ implies $a \succcurlyeq {\sim}x \succcurlyeq {\sim}a$. Thus, the interval $[{\sim}a, a]$ is closed with respect to the basic algebraic operators, and composes a subalgebra. □

Lemma 4.5. *Suppose X is a Boolean algebra. An object $a \in X$ is nonempty if and only if $a = \top$, and it is empty if and only if $a = \bot$. All other objects in X (besides $\top$ and $\bot$) are neither empty nor nonempty.*

Proof. Suppose a is nonempty. Applying Theorem 4.1, we obtain $a = x \vee {\sim}x$ for some $x \in X$. The law of excluded middle implies $a = \top$. Furthermore, notice that $\top = \top \vee \bot = \top \vee {\sim}\top$. Thus, $\top$ is nonempty. Likewise, suppose a is empty. Applying Theorem 4.1 again, we obtain $a = x \wedge {\sim}x$ for some $x \in X$. The law of noncontradiction implies $a = \bot$. Furthermore, notice that $\bot = \bot \wedge \top = \bot \wedge {\sim}\bot$. Thus, $\bot$ is empty. There is no object that is both empty and nonempty, because the top $\top$ is nonempty-only and the bottom $\bot$ is empty-only. Therefore, all other objects besides $\top$ and $\bot$ are neither empty nor nonempty. □

4.5 Exercises

(1) Suppose X is a de Morgan algebra. If $x \parallel {\sim}x$, carefully explain why the "tetralemma" generated by x, i. e., $\{x, {\sim}x, x \vee {\sim}x, x \wedge {\sim}x\}$ is a subalgebra of X. When is this configuration not a tetralemma?

(2) Suppose X is a de Morgan algebra. Demonstrate that $x \wedge {\sim}x = \bot$ if and only if $x \vee {\sim}x = \top$.

(3) Construct a de Morgan algebra X with nonempty-only element $a \in X$ such the equation $a = x \vee {\sim}x$ is unsolvable for some $x \in X$ such that $x \parallel {\sim}x$.

(4) Suppose X is a de Morgan algebra. Why is the top $\top$ always nonempty and the bottom $\bot$ always empty? If $\top$ is the only nonempty element, demonstrate that X is an ortholattice.

(5) Suppose X is a de Morgan algebra. Demonstrate that $x \vee \sim x \succ x \wedge \sim x$ if and only if $x \neq \sim x$.
(6) Verify that the Belnap logical lattice satisfies the de Morgan laws, and that the Belnap information lattice satisfies the "dual de Morgan laws":

$$\sim(x \vee y) = \sim x \vee \sim y;$$
$$\sim(x \wedge y) = \sim x \wedge \sim y.$$

(7) Suppose the de Morgan algebra X has two self-negating elements, say a and b. Demonstrate that $a \simeq b$. If $a \parallel b$, explain why $a \vee b$ is nonempty-only and $a \wedge b$ is empty-only.
(8) Suppose X is a de Morgan algebra such that the set of all nonempty elements is a principal up-set. Demonstrate that $x \vee \sim x \succcurlyeq y \wedge \sim y$ for every pair $\{x, y\}$ of elements in X.
(9) Suppose X is a de Morgan algebra. For a fixed element $x \in X$, define the map $f : [x \wedge \sim x, x] \to [\sim x, x \vee \sim x]$ by the condition that $f(z) = \sim z$. Verify that this map f is a well-defined dual isomorphism. Interpret the result in terms of the algebraic tetralemma (Figure 4.1).
(10) Suppose X is a finite de Morgan algebra. For a fixed element $x \in X$, explain why every path from $x \wedge \sim x$ to $x \vee \sim x$ that includes x is matched by another path of the same length from $x \wedge \sim x$ to $x \vee \sim x$ that includes $\sim x$.

5 Product algebras

One of the most important features of universal algebra is the natural way it handles the Cartesian product of multiple algebras. We develop this theory mainly for the product of two algebras, but the main results easily extend to higher dimensions. Suppose X and Y are two algebras. Then their *product* is defined by

$$X \times Y := \{(x, y) : x \in X, y \in Y\}.$$

We will use the same symbols for the logic operators on either factor. The logic operators on the product are defined component-wise:

$$\begin{aligned}(x_1, y_1) \vee (x_2, y_2) &:= (x_1 \vee x_2, y_1 \vee y_2),\\ (x_1, y_1) \wedge (x_2, y_2) &:= (x_1 \wedge x_2, y_1 \wedge y_2),\\ \sim(x, y) &:= (\sim x, \sim y).\end{aligned}$$

As a general guideline, if a certain algebraic condition holds for each factor, then it also holds for the product. In particular, the product of two lattices is a lattice, and the product of two de Morgan algebras is a de Morgan algebra. The main exception to this rule that we have studied so far is the concept of negative transitivity. The two components of a product tend to become entangled when we investigate incomparability.

A simple application of the product algebra is obtained by reanalyzing the four-way classification of the objects in the de Morgan algebra: emptiness, nonemptiness, both, and neither. We must first refine the four transcendental categories. Let $A := \{a \in X : a \preccurlyeq \sim a\}$ and $B := \{a \in X : a \succcurlyeq \sim a\}$ denote the two categories of emptiness and nonemptiness. Then the categories of "both" and "neither" are the subsets $A \cap B$ and $A' \cap B'$. We next replace A and B with the subsets $A \cap B'$ and $A' \cap B$. These subsets are readily interpreted as the categories of "emptiness only" and "nonemptiness only". We thus obtain $A \cap B' = \{a \in X : a \prec \sim a\}$ and $A' \cap B = \{a \in X : a \succ \sim a\}$. Therefore, the algebra X is partitioned into four basic categories: $A \cap B'$, $A' \cap B$, $A \cap B$, and $A' \cap B'$.

We next factor this partition into a two-attribute model, where the attributes are "emptiness" and "nonemptiness." Each attribute has two levels.

Category	Emptiness	Nonemptiness
Emptiness Only	A	B'
Nonemptiness Only	A'	B
Both	A	B
Neither	A'	B'

In essence, we are identifying each category as a bundle of attributes.

https://doi.org/10.1515/9783112235812-005

Category	Product
$A \cap B'$	(A, B')
$A' \cap B$	(A', B)
$A \cap B$	(A, B)
$A' \cap B'$	(A', B')

We think of the four possibilities as categories in order to assign the attributes to individual objects, but we must think of them as Cartesian products of attributes to construct the product algebra.

We next order the levels of each attribute, $A \succ A'$ and $B \succ B'$. These orders are a matter of convention. Indeed, the underlying lattice can be safely ignored at this stage of construction. Finally, we specify the two levels of each attribute to be opposites, so that $\sim A = A'$ and $\sim B = B'$. Thus, each attribute is modeled as a Boolean truth lattice. The product involution neatly organizes the four categories into two pairs of opposites, $\sim(A, B) = (A', B')$ and $\sim(A, B') = (A', B)$. The product algebra is isomorphic to a tetralemma with a Hasse diagram displayed in Figure 5.1. The semantics of this tetralemma are similar to the tetralemma of emptiness. This construction also reveals that the tetralemma is isomorphic to the product of two Boolean truth lattices. More importantly, it indicates the wide scope of the tetralemma as an instrument to investigate the general de Morgan algebra.

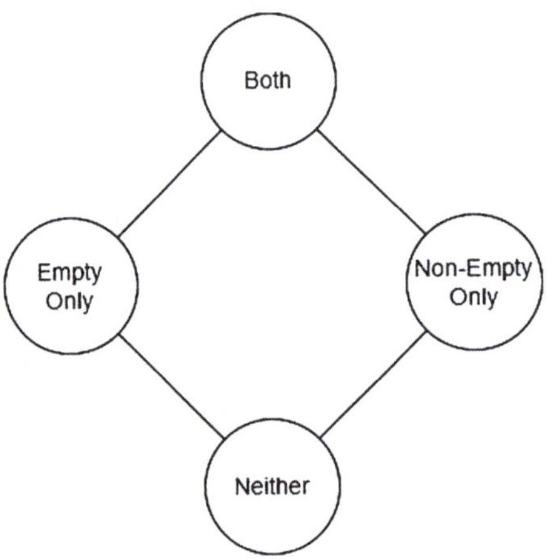

Figure 5.1: The tetralemma of the four transcendental categories.

5.1 Basic properties

Let us next examine how the four categories work with the objects in the product algebra. These properties are inherited by the product to the extent that they are algebraic forms.

Lemma 5.1. *Suppose $X \times Y$ is the product of the de Morgan algebras X and Y. The object $(x,y) \in X \times Y$ is nonempty if and only if x is nonempty and y is nonempty, (x,y) is empty if and only if x is empty and y is empty, and (x,y) is both (empty and nonempty) if and only if x is both and y is both.*

There remains the category of "neither emptiness nor nonemptiness." There are four basic reasons why an object $(x,y) \in X \times Y$ might belong to this category:
(1) $x \parallel \sim x$;
(2) $y \parallel \sim y$;
(3) $x \succ \sim x$ and $\sim y \succ y$;
(4) $\sim x \succ x$ and $y \succ \sim y$.

It is easy to verify $(x,y) \parallel (\sim x, \sim y)$ in all four cases, and this is the only reason for incomparability. This sometimes makes the analysis of "neither" in the product algebra more tedious than the analysis of the other categories. We also know from the one-dimensional analysis that $(x,y) \parallel (\sim x, \sim y)$ implies (x,y) and $(\sim x, \sim y)$ are the opposite medians of a tetralemma with nonempty top $(x \vee \sim x, y \vee \sim y)$ and empty bottom $(x \wedge \sim x, y \wedge \sim y)$. We next seek a converse to this implication in the context of the product algebra.

The equation we aim to solve in $X \times Y$ is $(a,b) = (x,y) \vee (\sim x, \sim y)$ with the constraint $(x,y) \parallel (\sim x, \sim y)$. One approach is to solve this equation in each dimension separately. Thus, we attempt to solve two equations: $a = x \vee \sim x$ with $x \parallel \sim x$; and $b = y \vee \sim y$ with $y \parallel \sim y$. However, the incomparability constraint that $(x,y) \parallel (\sim x, \sim y)$ is not equivalent to $x \parallel \sim x$ and $y \parallel \sim y$. Thus, we can miss some possible solutions this way. We must take into account all the possible ways that $(x,y) \parallel (\sim x, \sim y)$ can occur as we listed above. This task is not as difficult as it might seem.

The product order $\succcurlyeq$ on the product lattice $X \times Y$ is defined by the condition that $(x_1,y_1) \succcurlyeq (x_2,y_2)$ if and only if $x_1 \succcurlyeq x_2$ and $y_1 \succcurlyeq y_2$. Its asymmetric part is given by $(x_1,y_1) \succ (x_2,y_2)$ if and only if $(x_1,y_1) \succcurlyeq (x_2,y_2)$ and not $(x_1,y_1) \succcurlyeq (x_2,y_2)$. Consequently, equality can appear in one of the dimensions. On the other hand, the *strict product order* $\succ\!\succ$ on $X \times Y$ is defined by the condition that $(x_1,y_1) \succ\!\succ (x_2,y_2)$ if and only if $x_1 \succ x_2$ and $y_1 \succ y_2$.

Lemma 5.2. *Suppose $X \times Y$ is the product of the de Morgan algebras X and Y. If $(a,b) \succ\!\succ (\sim a, \sim b)$, then*

$$(a,b) = (a, \sim b) \vee (\sim a, b);$$
$$(\sim a, \sim b) = (a, \sim b) \wedge (\sim a, b).$$

Moreover, $(a, \sim b)$ and $(\sim a, b)$ are opposites such that $(a, \sim b) \parallel (\sim a, b)$.

In brief, we say that (a,b) is "strictly nonempty" and $(\sim a, \sim b)$ is "strictly empty" provided that $(a,b) \succ\!\succ (\sim a, \sim b)$. Thus, the result states that if (a,b) is strictly nonempty, it is the disjunction of two incomparable opposites. Moreover, $(\sim a, \sim b)$ is strictly empty, and it is the conjunction of the same two incomparable opposites. A tetralemma is thus found:

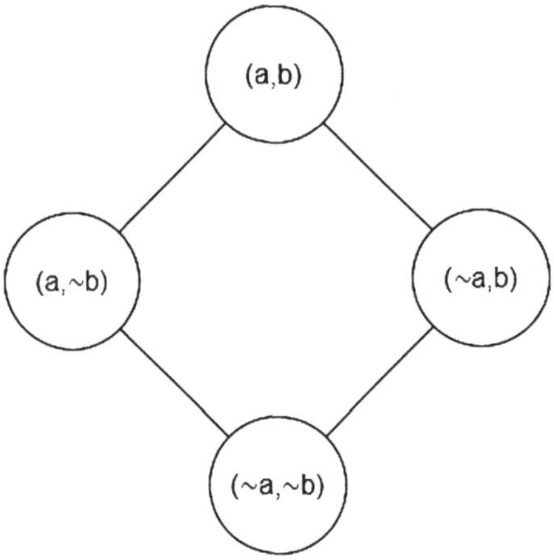

Figure 5.2: The tetralemma as a product algebra.

This is a subalgebra of the product algebra $X \times Y$. If neither X nor Y has a self-negating object, this is the end of the matter. Every nonempty object (a, b) is strictly nonempty and every empty object $(\sim a, \sim b)$ is strictly empty. Thus, Lemma 5.2 covers all the possibilities in this case. However, if either X or Y have a self-negating object, more work must be done.

Lemma 5.3. *Suppose a is a strictly nonempty object in the de Morgan algebra X, and b is a self-negating object in the de Morgan algebra Y. If the equation $(a, b) = (x, y) \vee (\sim x, \sim y)$ is solved by some (x, y) such that $(x, y) \parallel (\sim x, \sim y)$, then $y = b$ and $x \parallel \sim x$.*

In essence, this lemma reduces the two-dimensional problem back to a one-dimensional problem for the first component: Solve $a = x \vee \sim x$ with $x \parallel \sim x$.

5.2 Measures of order

We now explain how to extend the measures of order for the finite de Morgan algebra to the product algebra. We must assume that X and Y are finite lattices in order to compute canonical scores and standard scores. We begin with the canonical score. There is no elementary formula for the canonical score $\Phi(x, y)$ in terms of $\Phi(x)$ and $\Phi(y)$, but there is still a good workaround. We will use the same symbols, U and D, for the up- and down-sets in both X and Y.

Lemma 5.4. *If X and Y are lattices, then $U(x, y) = U(x) \times U(y)$ and $D(x, y) = D(x) \times D(y)$. If X and Y are finite lattices, then $u(x, y) = u(x)u(y)$ and $d(x, y) = d(x)d(y)$.*

This multiplicative property is all that is needed to compute the canonical score of a product from the measurements of its components,

$$\Phi(x,y) = \frac{d(x,y) - 1}{u(x,y) + d(x,y) - 2}.$$

We study the problem of determining the medians of the product algebra as an important area of application.

Corollary. *Suppose X and Y are finite lattices. If x is a canonical median of X and y is a canonical median of Y, then (x,y) is a canonical median of $X \times Y$. Moreover, if x is a canonical median of X, then (x,y) is a canonical median of $X \times Y$ if and only if y is a canonical median of Y.*

Lemma 5.5. *Suppose X and Y are finite de Morgan algebras. Then (x,y) is a canonical median of $X \times Y$ if and only if*

$$\frac{d(y)}{u(y)} = \frac{d(\sim x)}{u(\sim x)}.$$

Corollary. *Suppose X is a de Morgan algebra. The ordered pair $(x, \sim x)$ is a median of X^2. If the ordered pair (x,y) is a median of X^2, then $y \simeq \sim x$.*

This result guarantees that the product algebra X^2 has a canonical median regardless of whether X has one.

Example. Suppose X is specified by $1 > 1/2 > 0$, where $\{1, 0\}$ are opposites and $1/2$ is self-negated; and Y is specified by $1 > 0$, where $\{1, 0\}$ are opposites. (X is known as the "logic of paradox" and Y is the Boolean truth lattice.) The product algebra has the Hasse diagram displayed in Figure 5.3.

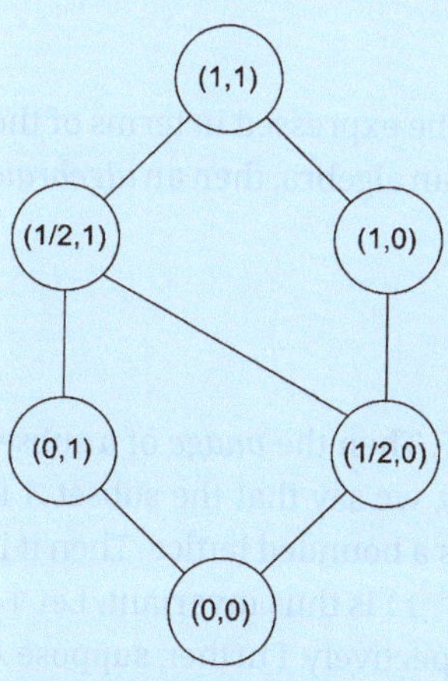

Figure 5.3: The product of the logic of paradox with the Boolean truth lattice.

Notice that the product algebra has no medians.

Clearly, if X and Y are modular lattices, then $X \times Y$ is a modular lattice. In the finite case, we aim to devise concise computational formulas for the standard score.

Lemma 5.6. *Suppose X is a finite modular lattice with maximum chain length m, and Y is a finite modular lattice with maximum chain length n. Then $X \times Y$ is a finite modular lattice with maximum chain length $m + n - 1$. Moreover, if $x \in X$ has rank i and $y \in Y$ has rank j, then $(x, y) \in X \times Y$ has rank $i + j - 1$.*

We next define $w_X := \frac{m-1}{m+n-2}$ and $w_Y := \frac{n-1}{m+n-2}$. Since $w_X + w_Y = 1$, we interpret them as the "relative weights" of X and Y in the product $X \times Y$.

Corollary. *Given the conditions of Lemma* 5.5, *for every $(x, y) \in X \times Y$,*

$$Z(x, y) = w_X Z(x) + w_Y Z(y).$$

When we compare the formulas for canonical scores and standard scores of objects in a product lattice, an interesting distinction arises. The canonical score is based on multiplication and the standard score is based on addition. This reflects the fact that the canonical score is a two-dimensional measure (width and length) of order, and the standard score is a one-dimensional measure (length).

5.3 Left–right symmetry

Suppose X is an algebra. An *automorphism* of X is an invertible map $f : X \to X$ that respects the algebraic operations on X. The map f is invertible if and only if it is one-to-one and onto. If X is a lattice, then a *lattice automorphism* f must satisfy the conditions:
(1) $f(x \vee y) = f(x) \vee f(y)$;
(2) $f(x \wedge y) = f(x) \wedge f(y)$.

This pair of conditions for a lattice automorphism can also be expressed in terms of the lattice order: $x \succcurlyeq y$ if and only if $f(x) \succcurlyeq f(y)$. If X is a de Morgan algebra, then an *algebraic automorphism* f must also satisfy a third condition:
(3) $f(\sim x) = \sim f(x)$.

In this case, we simply say that "f is an automorphism."

Suppose $f : X \to X$ is an automorphism of the lattice X. Then the *image* of a subset $A \subseteq X$ is defined by $f(A) := \{f(x) : x \in A\}$. More generally, we say that the subset A is *invariant* provided that $f(A) = A$. For example, suppose X is a bounded lattice. Then it is easy to verify that $f(\top) = \top$ and $f(\bot) = \bot$. The sublattice $\{\top, \bot\}$ is thus invariant. Let $\top_A$ and $\bot_A$ denote the top and bottom of the subalgebra A, respectively. Further, suppose X is a de Morgan algebra. Since $\top_A \succ \bot_A$ and $\sim \top_A = \bot_A$, we deduce that $\top_A$ is nonempty and $\bot_A$ is empty. Furthermore, $f(\top_A) = \top_A$ and $f(\bot_A) = \bot_A$. The top $\top_A$ and bottom $\bot_A$ of A are thus "fixed points" of the automorphism f.

Theorem 5.7. *Suppose $f : X \to X$ is an automorphism of the de Morgan algebra X. If the subalgebra A is invariant, then the interval subalgebra $[\bot_A, \top_A]$ is invariant.*

Corollary. *Suppose $f : X \to X$ is an automorphism of the de Morgan algebra X. The interval subalgebra $[\sim a, a]$, where $a \succ \sim a$, is invariant if and only if the endpoint a is a fixed point, i. e., $f(a) = a$.*

This theorem reduces the problem of finding an invariant subalgebra to the problem of finding a fixed point $f(a) = a$ such that $a \succ \sim a$. It is easy to verify that $\sim a$ is a fixed point, i. e., $f(\sim a) = \sim a$, if and only if a is a fixed point. Only in this case is the interval subalgebra $[\sim a, a]$ invariant. Moreover, every invariant subalgebra A of X must also be an invariant subalgebra of $[\bot_A, \top_A]$.

Lemma 5.8. *Suppose X is a finite lattice. If the map $f : X \to X$ is a lattice automorphism, then $f(x) \simeq x$ for every $x \in X$.*

By definition, we say that an automorphism $f : X \to X$ is *idempotent* provided that $f^2(x) = x$ for every $x \in X$. As a special case, we say that X exhibits *left–right symmetry* provided that there exists an idempotent lattice automorphism on X.

Example. Consider the six-valued de Morgan algebra X displayed in Figure 5.4.

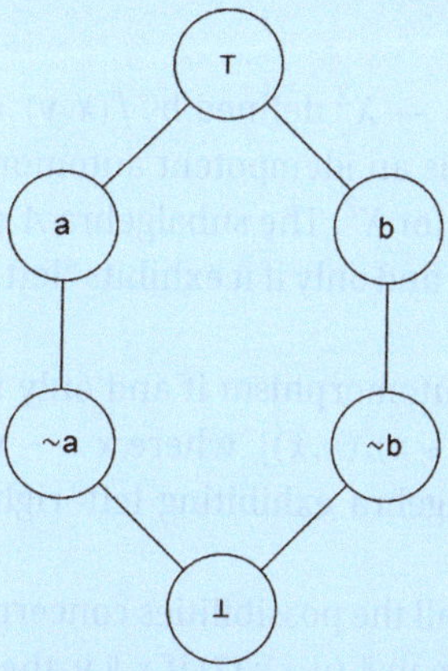

Figure 5.4: A Hasse diagram with left–right symmetry.

Define a map $f : X \to X$ by the following table:

x	$\top$	a	b	$\sim a$	$\sim b$	$\bot$
$f(x)$	$\top$	b	a	$\sim b$	$\sim a$	$\bot$

In essence, the map $f : X \to X$ exchanges values with the same rank. Thus, it is a lattice automorphism. Since it also preserves the involution, it is an algebraic automorphism. The only fixed points are the top $\top$ and the bottom $\bot$. Since we compute $f^2(x) = x$ for every $x \in X$, the algebra X exhibits left–right symmetry.

Example. The Belnap information lattice is represented by the Hasse diagram in Figure 5.5.

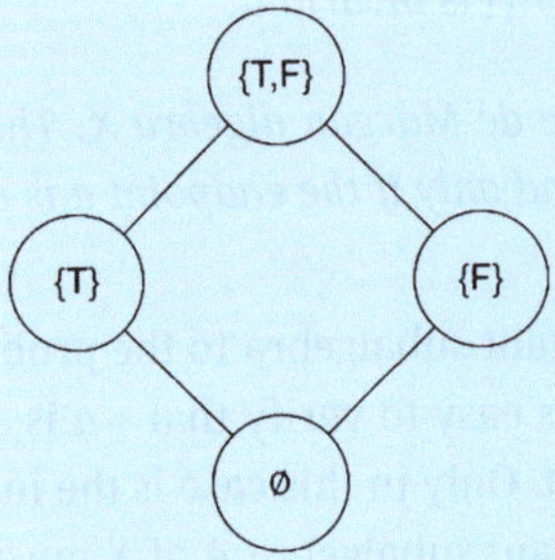

Figure 5.5: The Belnap information lattice.

The Belnap "negation" operator $f : X \to X$ is defined element-wise upon the power set representation, which is summarized in the following table:

x	$\{T,F\}$	$\{T\}$	$\{F\}$	$\emptyset$
$f(x)$	$\{T,F\}$	$\{F\}$	$\{T\}$	$\emptyset$

It is easy to verify f is an idempotent automorphism. It thus displays left–right symmetry. The Belnap information lattice is not a de Morgan algebra with respect to f, because it violates the de Morgan laws.

This theory can also be applied to the automorphism $f : X^2 \to X^2$ defined by $f(x,y) = (y,x)$, where $X^2 := X \times X$ is the product algebra. This map is an idempotent automorphism, which we identify as the "left–right automorphism" for X^2. The subalgebra $A \subseteq X^2$ is invariant with respect to the left–right automorphism if and only if it exhibits "left–right symmetry": $(x,y) \in A$ if and only if $(y,x) \in A$.

An object $(x,y) \in X^2$ is a fixed point of the left–right automorphism if and only if $x = y$. An interval subalgebra of X^2 thus has the form $[(\sim x, \sim x), (x,x)]$, where $x \succ \sim x$. Each subalgebra of such an interval is an invariant subalgebra exhibiting left–right symmetry.

We can also verify that $f(x,y) \simeq (x,y)$. We simply cover all the possibilities concerning the relation between x and y: (1) If $x \succ y$ or $y \succ x$, then $(x,y) \parallel (y,x)$; (2) if $x \parallel y$, then $(x,y) \parallel (y,x)$; and (3) if $x = y$, then $(x,y) = (y,x)$. Therefore, $(y,x) \simeq (x,y)$ no matter what order prevails between x and y.

In the finite case, we can say a little more. First, Lemma 5.4 clearly implies that $u(x,y) = u(y,x)$ and $d(x,y) = d(y,x)$. Thus, we obtain $\Phi(x,y) = \Phi(y,x)$. Second, if X is modular, then the joint ranks of (x,y) and (y,x) are the same. This result follows from Lemma 5.5. Thus, we obtain $Z(x,y) = Z(y,x)$.

Example. We next study the product of two tetralemmas as an application of this theory. The labels of the tetralemma X in Figure 5.6 are abbreviated as T, F, B, and N to facilitate the construction.

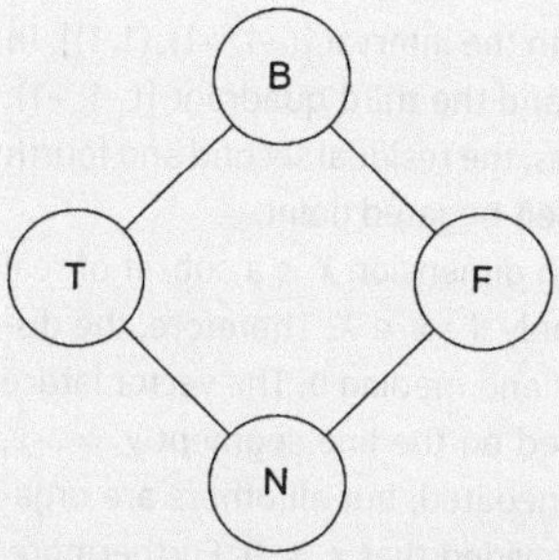

Figure 5.6: A tetralemma with generic labels.

The product X^2 of two tetralemmas has 16 truth-values. This is a relatively large, but tractable number. Accordingly, the Hasse diagram in Figure 5.7 is somewhat unwieldy. The theory serves to identify various subalgebras. The six medians are organized into three pairs of incomparable opposites:

$$\{(B,N),(N,B)\}, \quad \{(T,T),(F,F)\}, \quad \text{and} \quad \{(T,F),(F,T)\}.$$

Each of these pairs forms a tetralemma with top (B,B) and bottom (N,N). The product exhibits left–right symmetry. There are four fixed points of the left–right automorphism: (B,B), (T,T), (F,F), and (N,N). However, only two them form a proper interval, i. e. $[(N,N),(B,B)]$, but this interval includes the entire product algebra X^2.

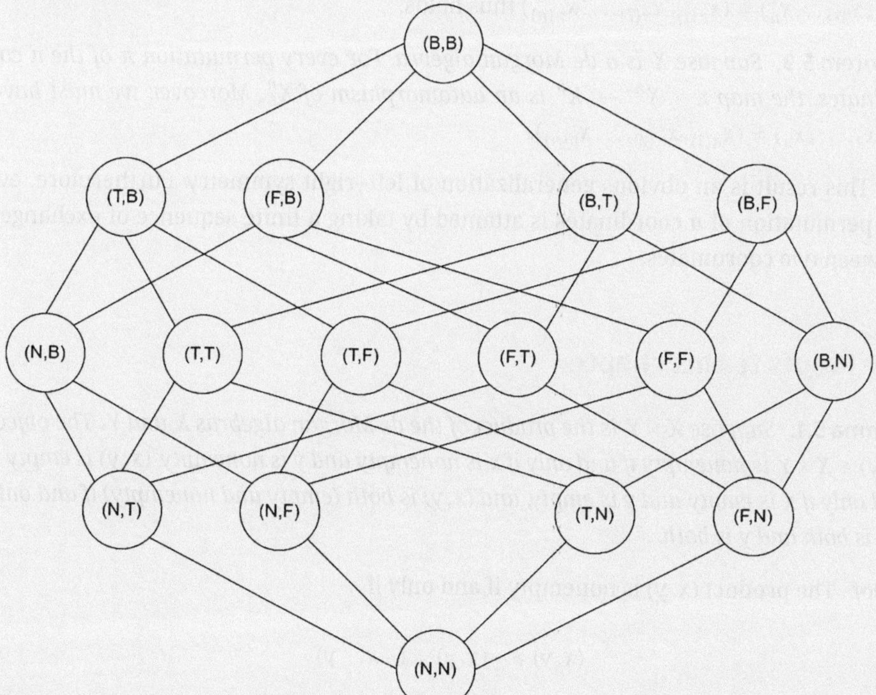

Figure 5.7: The product of two tetralemmas.

Example. Consider the vector lattice of ordered pairs of real numbers in the interval $[(-1,-1),(1,1)]$. In general, the first quadrant $[(0,0),(1,1)]$ consists of nonempty points, and the third quadrant $[(-1,-1),(0,0)]$ consists of empty points. Excluding the vertical and horizontal axes, the residual second and fourth quadrants consist of incomparable points. The origin $(0,0)$ is the only self-negated point.

We can discretize this vector lattice in various ways. Suppose each dimension X is a subset of real numbers in the interval $[-1,1]$ that includes $\{0,\pm 1\}$, and $x \in X$ if and only if $-x \in X$. Therefore, the discretized interval X is isomorphic to a fuzzy lattice with top 1, bottom -1, and median 0. The vector lattice $[(-1,-1),(1,1)]$ is discretized as X^2. In general, the medians are located on the line segment $y = -x$, i. e., (x,y) is a median if and only if $y = -x$. The median $(0,0)$ is self-negated, but all others are organized into incomparable pairs, i. e. $(-x,x)$ is the opposite of $(x,-x)$ provided that $x \neq 0$. Furthermore, the interval subalgebras have the generic form $[(-x,-y),(x,y)]$, where $x > -x$ and $y > -y$. Left–right symmetric intervals of the generic form $[(-x,-x),(x,x)]$ are invariant with respect to the left–right automorphism.

Although we have not discussed the product of more than two factors, it is clear that the theory of product algebras naturally extends this way. The way that left–right symmetry generalizes is in terms of permutations of the coordinates. For example, (z,y,x) is a permutation of (x,y,z). In formal terms, a permutation is an invertible (i. e., one-to-one and onto) map $\pi : \{1,2,3,\dots,n\} \to \{1,2,3,\dots,n\}$. Using "wrong but clear" notation, we also designate the map permuting the coordinates as $\pi : X^n \to X^n$. The identity $\pi(x_1,x_2,\dots,x_n) = (x_{\pi(1)},x_{\pi(2)},\dots x_{\pi(n)})$ thus holds.

Theorem 5.9. *Suppose X is a de Morgan algebra. For every permutation π of the n coordinates, the map $\pi : X^n \to X^n$ is an automorphism of X^n. Moreover, we must have $(x_1,x_2,\dots,x_n) \simeq (x_{\pi(1)},x_{\pi(2)},\dots x_{\pi(n)})$.*

This result is an obvious generalization of left–right symmetry. Furthermore, every permutation of n coordinates is attained by taking a finite sequence of exchanges between two coordinates.

5.4 Proofs to this chapter

Lemma 5.1. *Suppose $X \times Y$ is the product of the de Morgan algebras X and Y. The object $(x,y) \in X \times Y$ is nonempty if and only if x is nonempty and y is nonempty, (x,y) is empty if and only if x is empty and y is empty, and (x,y) is both (empty and nonempty) if and only if x is both and y is both.*

Proof. The product (x,y) is nonempty if and only if

$$(x,y) \succcurlyeq \sim(x,y) = (\sim x, \sim y).$$

But $(x,y) \succcurlyeq (\sim x,\sim y)$ if and only if $x \succcurlyeq \sim x$, i. e., x is nonempty, and $y \succcurlyeq \sim y$, i. e., y is nonempty. The other parts of the lemma are proved similarly. □

Lemma 5.2. *Suppose $X \times Y$ is the product of the de Morgan algebras X and Y. If $(a,b) \gg (\sim a, \sim b)$, then*

$$(a,b) = (a, \sim b) \vee (\sim a, b);$$
$$(\sim a, \sim b) = (a, \sim b) \wedge (\sim a, b).$$

Moreover, $(a, \sim b)$ and $(\sim a, b)$ are opposites such that $(a, \sim b) \parallel (\sim a, b)$.

Proof. Since $(a,b) \gg (\sim a, \sim b)$, we have $a \succ \sim a$ and $b \succ \sim b$. Thus, the above equations are identities. Moreover, $(a, \sim b) \parallel (\sim a, b)$, because the components are ordered in opposite directions. □

Lemma 5.3. *Suppose a is a strictly nonempty object in the de Morgan algebra X, and b is a self-negating object in the de Morgan algebra Y. If the equation $(a,b) = (x,y) \vee (\sim x, \sim y)$ is solved by some (x,y) such that $(x,y) \parallel (\sim x, \sim y)$, then $y = b$ and $x \parallel \sim x$.*

Proof. Assume $(a,b) = (x \vee \sim x, y \vee \sim y)$. Then $b = y \vee \sim y \succcurlyeq y \succcurlyeq y \wedge \sim y = \sim b$ and $b = \sim b$ implies $y = b$. Since $y = \sim y$, the relation $(x,y) \parallel (\sim x, \sim y)$ requires that $x \parallel \sim x$. □

Lemma 5.4. *If X and Y are de Morgan algebras, then $U(x,y) = U(x) \times U(y)$ and $D(x,y) = D(x) \times D(y)$. If X and Y are finite de Morgan algebras, then $u(x,y) = u(x)u(y)$ and $d(x,y) = d(x)d(y)$.*

Proof. Clearly, $(x_1, y_1) \in U(x,y)$ if and only if $(x_1, y_1) \succcurlyeq (x,y)$. The latter condition holds if and only if $x_1 \succcurlyeq x$ and $y_1 \succcurlyeq y$, i. e., $x_1 \in U(x)$ and $y_1 \in U(y)$. Thus, we obtain $U(x,y) = U(x) \times U(y)$. The proof of $D(x,y) = D(x) \times D(y)$ is by duality. For every $y_1 \in U(y)$, we find precisely $u(x)$ values of X, say x_1, such that $(x_1, y_1) \in U(x,y)$. Thus, we obtain $u(x,y) = u(x)u(y)$. A similar proof holds for $d(x,y) = d(x)d(y)$. □

Corollary. *Suppose X and Y are finite lattices. If x is a canonical median of X and y is a canonical median of Y, then (x,y) is a canonical median of $X \times Y$. Moreover, if x is a canonical median of X, then (x,y) is a canonical median of $X \times Y$ if and only if y is a canonical median of Y.*

Proof. For the first part, notice that $u(x) = d(x)$ and $u(y) = d(y)$ imply

$$u(x,y) = u(x)u(y) = d(x)d(y) = d(x,y).$$

For the second part, notice that $u(x) = d(x)$ implies $u(x)u(y) = d(x)d(y)$ if and only if $u(y) = d(y)$. □

Lemma 5.5. *Suppose X and Y are finite de Morgan algebras. Then (x,y) is a canonical median of $X \times Y$ if and only if*

$$\frac{d(y)}{u(y)} = \frac{d(\sim x)}{u(\sim x)}.$$

Proof. One has $\Phi(x,y) = 1/2$ if and only if $u(x)u(y) = d(x)d(y)$. But $u(x) = d(\sim x)$ and $d(x) = u(\sim x)$ by virtue of polarity. The result immediately follows. □

Corollary. *Suppose X is a de Morgan algebra. The ordered pair $(x, \sim x)$ is a median of X^2. If the ordered pair (x,y) is a median of X^2, then $y \simeq \sim x$.*

Proof. For the first part, we simply specify $y = \sim x$ in Lemma 5.5. The proof of the second part is by contradiction. Contrary to hypothesis, assume $y \not\simeq \sim x$, so that $y \succ \sim x$ or $\sim x \succ y$. In the former case, notice that $u(y) < u(\sim x)$ and $d(y) > d(\sim x)$ imply

$$u(x,y) < u(x, \sim x) = d(x, \sim x) < d(x,y).$$

Thus, (x,y) is not a median. In the latter case, notice that $u(y) > u(\sim x)$ and $d(y) < d(\sim x)$ imply

$$u(x,y) > u(x, \sim x) = d(x, \sim x) > d(x,y).$$

Thus, (x,y) is not a median. The result immediately follows. □

Lemma 5.6. *Suppose X is a finite modular lattice with maximum chain length m, and Y is a finite modular lattice with maximum chain length n. Then $X \times Y$ is a finite modular lattice with maximum chain length $m + n - 1$. Moreover, if $x \in X$ has rank i and $y \in Y$ has rank j, then $(x,y) \in X \times Y$ has rank $i + j - 1$.*

Proof. Since the modular law is constructed from disjunctions and conjunctions, it is easy to verify that a product lattice $X \times Y$ satisfies the modular law if and only if each factor, i. e., X or Y, satisfies the modular law. Next we let α denote a maximal chain in X from $\perp$ to $\top$, and let β denote a maximal chain in Y from $\perp$ to $\top$. Then we construct a chain in $X \times Y$ from the bottom $(\perp, \perp)$ to the top $(\top, \top)$ as follows: We first follow the chain $\{\perp\} \times \beta$ from $(\perp, \perp)$ to $(\perp, \top)$. Then we switch to the chain $\alpha \times \{\top\}$ from $(\perp, \top)$ to $(\top, \top)$. This chain has length $m + n - 1$. Moreover, it is a maximal chain, because α and β are maximal chains. The formula for the rank of (x,y) is derived similarly. □

Corollary. *Given the conditions of Lemma 5.5, for every $(x,y) \in X \times Y$,*

$$Z(x,y) = w_X Z(x) + w_Y Z(y).$$

Proof. The standard score is given by

$$Z(x,y) = \frac{(i + j - 1) - 1}{(m + n - 1) - 1} = \frac{i - 1}{m + n - 2} + \frac{j - 1}{m + n - 2}.$$

It is easy to arrange the two terms so that one is $w_X Z(x)$ and the other is $w_Y Z(y)$. □

Theorem 5.7. *Suppose $f : X \to X$ is an automorphism of the de Morgan algebra X. If the subalgebra A is invariant, then the interval subalgebra $[\perp_A, \top_A]$ is invariant.*

Proof. Since $x \succcurlyeq \perp_A$ for every $x \in A$, we obtain $f(x) \succcurlyeq f(\perp_A)$. Since A is an invariant subalgebra, $f : A \to A$ is onto. Thus, $x \succcurlyeq f(\perp_A)$ for every $x \in A$, which implies $f(\perp_A) = \perp_A$. A similar argument shows that $f(\top_A) = \top_A$. Further, notice that $\top_A \succcurlyeq x \succcurlyeq \perp_A$ implies $f(\top_A) \succcurlyeq f(x) \succcurlyeq f(\perp_A)$, which simplifies to $\top_A \succcurlyeq f(x) \succcurlyeq \perp_A$. Thus, the interval subalgebra $[\perp_A, \top_A]$ is invariant. □

Corollary. *Suppose $f : X \to X$ is an automorphism of the de Morgan algebra X. The interval subalgebra $[\sim a, a]$, where $a \succ \sim a$, is invariant if and only if the endpoint a is a fixed point, i. e., $f(a) = a$.*

Proof. Suppose the interval subalgebra $[\sim a, a]$ is invariant, where $a \succ \sim a$. The proof to the previous theorem verifies that its top a is a fixed point, i. e., $f(a) = a$. Conversely, assume $f(a) = a$. It follows that $f(\sim a) = \sim f(a) = \sim a$. We next select $x \in [\sim a, a]$. Then $a \succcurlyeq x \succcurlyeq \sim a$ implies $f(a) \succcurlyeq f(x) \succcurlyeq f(\sim a)$, which simplifies to $a \succcurlyeq f(x) \succcurlyeq \sim a$. Therefore, the interval $[\sim a, a]$ is invariant. □

Lemma 5.8. *Suppose X is a finite lattice. If the map $f : X \to X$ is a lattice automorphism, then $f(x) \simeq x$ for every $x \in X$.*

Proof. Let $a \in X$. Since f is an automorphism, it preserves order, i. e., $x \succcurlyeq y$ if and only if $f(x) \succcurlyeq f(y)$. The identity $f[U(a)] = U[f(a)]$ implies $u[f(a)] = u(a)$. Assume $f(a) \succ a$. It follows that $U[f(a)] \subseteq U(a)$ and that $u[f(a)] < u(a)$. This contradiction eliminates the possibility that $f(a) \succ a$. Furthermore, the dual identity $f[D(a)] = D[f(a)]$ implies $d[f(a)] = d(a)$. Assume $a \succ f(a)$. Then $D[f(a)] \subseteq D(a)$ and $d[f(a)] < d(a)$. This contradiction eliminates the possibility $a \succ f(a)$. Since the cases $f(a) \succ a$ and $a \succ f(a)$ have been eliminated, the conclusion is $f(a) \simeq a$. □

Theorem 5.9. *Suppose X is a de Morgan algebra. For every permutation π of the n coordinates, the map $\pi : X^n \to X^n$ is an automorphism of X^n. Moreover, we must have $(x_1, x_2, \dots, x_n) \simeq (x_{\pi(1)}, x_{\pi(2)}, \dots x_{\pi(n)})$.*

Proof. Since the lattice operations are defined component-wise, they are clearly preserved by coordinate permutations. Thus, π defines an automorphism. For the next part, notice that

$$u(x_1, x_2, \dots, x_n) = u(x_{\pi(1)}, x_{\pi(2)}, \dots x_{\pi(n)})$$

and

$$d(x_1, x_2, \dots, x_n) = d(x_{\pi(1)}, x_{\pi(2)}, \dots x_{\pi(n)})$$

by virtue of Lemma 5.4. (This lemma is generalized by induction on n.) Contrary to hypothesis, assume $(x_1, x_2, \dots, x_n) \succ (x_{\pi(1)}, x_{\pi(2)}, \dots x_{\pi(n)})$. Then it follows that

$$u(x_1, x_2, \dots, x_n) < u(x_{\pi(1)}, x_{\pi(2)}, \dots x_{\pi(n)})$$

and

$$d(x_1, x_2, \dots, x_n) > d(x_{\pi(1)}, x_{\pi(2)}, \dots x_{\pi(n)}).$$

This contradicts their equality. Likewise, the assumption that

$$(x_1, x_2, \dots, x_n) \prec (x_{\pi(1)}, x_{\pi(2)}, \dots x_{\pi(n)})$$

leads to another contradiction. Eliminating these two possibilities, we obtain $(x_1, x_2, \dots, x_n) \simeq (x_{\pi(1)}, x_{\pi(2)}, \dots x_{\pi(n)})$. □

5.5 Exercises

(1) Suppose $X \times Y$ is the product of two de Morgan algebras, X and Y. Verify that $(x, y) \in X \times Y$ is nonempty-only if and only if (i) both x and y are nonempty and (ii) either x or y is nonempty-only.

(2) Suppose $f : X \to X$ is an automorphism onto the de Morgan algebra X. If the element $a \in X$ is the unique self-negating of X, verify that it is a fixed point of f, i. e., $f(a) = a$. More broadly, verify that the set of all self-negating elements in X is an invariant subset.

(3) Suppose $f : X \to X$ is an automorphism onto the de Morgan algebra X. Define $f^2 = f \circ f$. Demonstrate that $f^2(\sim x) = \sim f^2(x)$.

(4) Suppose $X \times Y$ is the product of two de Morgan algebras, X and Y. Demonstrate that the middle region

$$M_{X \times Y} := \{(x, y) \in X \times Y : (x, y) \simeq (\sim x, \sim y)\}$$

is an invariant subset of the left–right automorphism.

(5) Generalize Lemma 5.6 to the product of three finite modular lattices, e. g., $X \times Y \times Z$.

(6) Define a measure Ψ on the de Morgan algebra X by $\Psi(x) := d(x)/u(x)$. Suppose $X \times Y \times Z$ is the product of three finite de Morgan algebras. Verify that

$$\Psi(x, y, z) = \Psi(x)\Psi(y)\Psi(z)$$

for every $(x, y, z) \in X \times Y \times Z$.

(7) Suppose $X \times Y \times Z$ is the product of three finite de Morgan algebras. Verify that $(x, y, z) \in X \times Y \times Z$ is a canonical median if and only if

$$u(x)u(y)u(z) = d(x)d(y)d(z).$$

(8) Suppose X is a de Morgan algebra. Verify that the following four ordered pairs compose a tetralemma in X^2 for every $x \in X$ such that $x \neq \sim x$:

$$\{(x, \sim x), (\sim x, x), (x \vee \sim x, x \vee \sim x), (x \wedge \sim x, x \wedge \sim x)\}.$$

Demonstrate that this tetralemma is an invariant subalgebra of X^2 with respect to the left–right automorphism.

(9) Suppose X is a finite modular de Morgan algebra. Express the diamond equation for X^2 in terms of the diamond equation for X.

(10) Suppose X is the Belnap logical lattice. Compute all the fixed points of the left–right automorphism for X^2. Identify the empty and nonempty fixed points.

6 Fuzzy logic

Fuzzy logic is the most widely known alternative to conventional logic, which is essentially the logic of sets. The foremost pioneer of fuzzy logic is Zadeh [10], but the basic ideas can be traced to the early twentieth century studies of Tarski and Łukasiewicz [11]. It is traditionally constructed with a continuum of numeric truth-values corresponding to the unit interval $[0,1]$ of real numbers. We will use the symbol Ψ for the truth function that assigns these numbers to the de Morgan algebra X of objects. However, the objects in fuzzy logic are often just the numbers. The *fuzzy truth lattice* satisfies the following algebraic properties:

(1) $\Psi(\top) = 1$ and $\Psi(\bot) = 0$;
(2) $\Psi(x \vee y) = \max\{\Psi(x), \Psi(y)\}$;
(3) $\Psi(x \wedge y) = \min\{\Psi(x), \Psi(y)\}$;
(4) $\Psi(\sim x) = 1 - \Psi(x)$.

It is well-known that fuzzy logic is a special case of the de Morgan algebra. (In fact, it is a Kleene algebra.)

An interval subalgebra is an interval of the form $[a, 1-a]$, where a is a number such that $0 \leq a < 1/2$. The number a is the bottom $\bot_A$ and the number $1-a$ is the top $\top_A$ of the interval subalgebra A. More generally, a subalgebra A consists of a closed subset of $[0,1]$ such that $x \in A$ if and only if $(1-x) \in A$. A subalgebra is not necessarily a continuous interval of real numbers.

6.1 Discrete fuzzy logic

A *discrete fuzzy logic* is a finite set of numbers in the unit interval that satisfy the four fuzzy logic axioms. We also follow the convention that a discrete fuzzy logic must have at least three objects. This eliminates the Boolean truth lattice from consideration. Since the top and bottom are interpreted as "not fuzzy," we need at least one fuzzy object between them. Such a system is clearly a subalgebra of the fuzzy logic continuum. Conditions (2) and (3) of the fuzzy logic axioms are always satisfied by a subset of real numbers. Thus, we only need to check conditions (1) and (4) to verify that a finite set of numbers in the unit interval is a discrete fuzzy logic.

The ontological interpretation of fuzziness is thought-provoking. Suppose we imagine the top as truth, nonemptiness, or existence; and the bottom as falsehood, emptiness, or nonexistence. Then a fuzzy level is a mixture of truth and falsehood, emptiness and nonemptiness, or existence and nonexistence. The algebraic interpretation is straightforward: All levels of fuzziness greater than 1/2 are nonempty; all levels of fuzziness less than 1/2 are empty; and the fuzzy level of 1/2 is both empty and nonempty. The measure of fuzziness thus refines the relative extent of emptiness versus nonemptiness.

https://doi.org/10.1515/9783112235812-006

The list of fuzzy logic axioms is similar to the lists of algebraic properties of either the canonical score or the standard score. The critical difference is that the equalities in conditions (2) and (3) were merely inequalities before. Moreover, we constructed the canonical score and standard score only for finite algebras. We next investigate when these truth functions can be strengthened to the level of a discrete fuzzy logic.

Theorem 6.1. *Suppose X is a finite de Morgan algebra. Then X is isomorphic to a discrete fuzzy logic if and only if the lattice order on X is total. If the lattice order is total, then the standard score and canonical score are numerically identical measurements, and either one of them represents the discrete fuzzy logic.*

A unique feature of the discrete fuzzy logic is that the involution can be constructed directly from the lattice order. The basic idea is summarized by the next result.

Lemma 6.2. *Suppose X is a finite total lattice. If* $Z(x) = r$, *where* $r \in [0,1]$, *then there exists a unique object* $y \in X$ *such that* $Z(y) = 1 - r$.

The involution $\sim$ is constructed by the condition that $\sim x$ is the unique object y such that $Z(y) = 1 - Z(x)$. In practical terms, we pair the extremes of $\bot$ and $\top$ as opposites, the next pair of second most extreme objects as opposites, and march inward in this fashion until we reach the middle. If the lattice has an odd number of objects, then the one in the middle has no opposite and is designated as a self-negated object. The laws of double negation and polarity are clearly satisfied by this construction.

Example. The simplest example of a discrete fuzzy logic is the logic of paradox, which we denote as $\mathcal{P}$. Figure 6.1 is the Hasse diagram for this system.

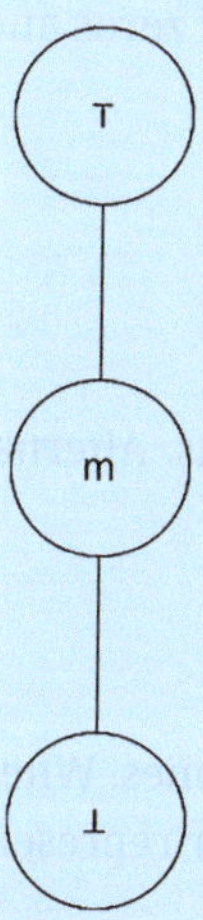

Figure 6.1: The logic of paradox.

The object m in the middle is self-negating, and the top $\top$ and bottom $\bot$ are opposites. The canonical (or standard) scores are given by $\Phi(\top) = 1$, $\Phi(\bot) = 0$, and $\Phi(m) = 1/2$. In terms of the four transcendental categories, the top is nonempty, the bottom is empty, and the middle is both empty and nonempty. The logic of paradox is thus a plausible model of T'ien-tai's three-truth theory in Ekayana Buddhism: The Truth of Emptiness, the Truth of Nonemptiness; and the Truth of the Middle Way. "Nonemptiness" in this context refers to conventional reality. The only proper subalgebra of the logic of paradox is the Boolean truth lattice, i. e., $\{\bot, \top\}$. The singleton $\{m\}$ can also be regarded as a trivial subalgebra in as much as one element m serves as the top, bottom, and its own involution.

6.2 Power sets

We next review a well-known representation theory of the power set as a Boolean algebra. This exercise is a preparation for fuzzy set theory in the next section. Let Ω denote a finite set, i. e.,

$$\Omega := \{a_1, a_2, \ldots, a_m\},$$

where each object a_i is an *element* of the superset Ω. It is an "element" in the sense that the singleton $\{a_i\}$ is an irreducible subset or "atom" of the universe Ω. By definition, the *power set* 2^Ω is the category of all subsets of Ω, including the empty set $\emptyset$ and the superset Ω.

The power set 2^Ω is a lattice, when we specify disjunction as union and conjunction as intersection. The distributive laws are easy to verify. Moreover, the lattice order is the same as set-inclusion. The top is Ω and the bottom is $\emptyset$. When we specify the involution as set-theoretic complementation, the power set is a de Morgan algebra. Furthermore, the laws of excluded middle and noncontradiction are satisfied. Therefore, the power set 2^Ω is a Boolean algebra.

We next review a numeric representation of the objects in 2^Ω in terms of zeros and ones. The indicator variable $1_A : \Omega \to \{0, 1\}$ is defined by

$$1_A[x] = \begin{cases} 1 & \text{if } x \in A, \\ 0 & \text{if } x \notin A \end{cases}$$

for every $A \in 2^\Omega$. Notice that these indicator variables represent the subsets. Alternatively, we identify the subset A with its vector representation:

$$A \leftrightarrow (1_A[a_1], 1_A[a_2], \ldots, 1_a[a_m]).$$

Thus, the subset A has a unique representation as an m-vector of zeros and ones. When we apply the first two operations (i. e., union and intersection) to their vector representations, they correspond to the product lattice on m-vectors:

$$(x_1, x_2, \ldots, x_m) \vee (y_1, y_2, \ldots, y_m) = (x_1 \vee y_1, x_2 \vee y_2, \ldots x_m \vee y_m);$$

$$(x_1, x_2, \ldots, x_m) \wedge (y_1, y_2, \ldots, y_m) = (x_1 \wedge y_1, x_2 \wedge y_2, \ldots x_m \wedge y_m).$$

The third formula for complementation is restricted to the unit interval $[0,1]$. Thus, the vector representation is restricted to $[0,1]^m$. Then the formula for complementation corresponds to the product involution:

$$\sim(x_1, x_2, \ldots, x_m) = (1 - x_1, 1 - x_2, \ldots, 1 - x_m).$$

Finally, one can verify that set inclusion $A \subseteq B$ corresponds to m-vector order:

$$(x_1, x_2, \ldots, x_m) \leq (y_1, y_2, \ldots y_m).$$

This order is the same as the product order on $[0,1]^m$. This derivation explains why the power set 2^Ω is isomorphic to the product algebra $\mathcal{B}^m$, where $\mathcal{B}$ is the Boolean truth lattice (i. e., $1 \succ 0$).

Example. The tetralemma is an interesting and important example of this representation theory. Let $\Omega = \{a_1, a_2\}$. Then the subsets of Ω are represented by 2-vectors of zeros and ones:

Subset	$\{a_1, a_2\}$	$\{a_1\}$	$\{a_2\}$	$\{\,\}$
Vector	(1,1)	(1,0)	(0,1)	(0,0)

Set-inclusion obviously corresponds to the vector order. The nodes of the Hasse diagram can be represented either as subsets or as vectors. Figure 6.2 labels them as ordered pairs of zeros and ones.

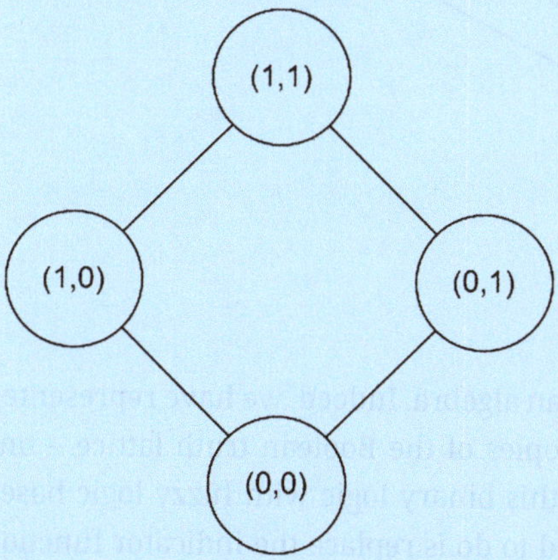

Figure 6.2: The tetralemma of all subsets of a 2-element set.

No matter how we label the four nodes of the tetralemma, the top is nonempty, the bottom is empty, and the middle is neither empty nor nonempty.

We can also use these digital labels to encode the elements of the product of two tetralemmas as four-tuples of zeros and ones. The Hasse diagram displayed in Figure 6.3 is the result. The underlying lattice has been studied in computer science applications as the product of the Belnap logical lattice with the Belnap information lattice [12]. The digital representation is convenient, because it makes both the lattice order and the location of opposites transparent. Of course, this algebra is isomorphic to the one displayed in Figure 5.7.

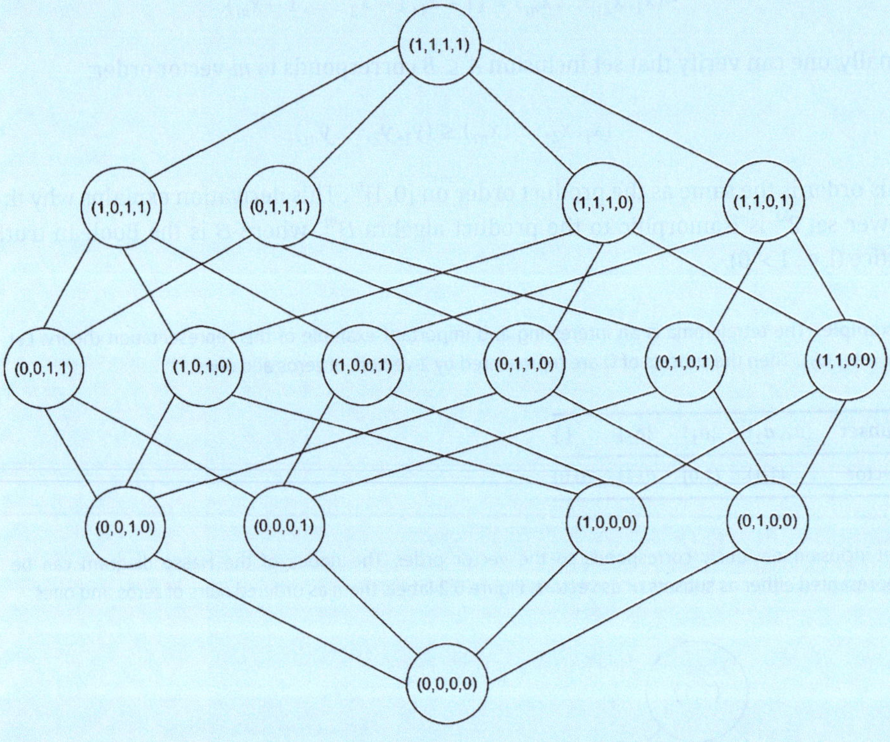

Figure 6.3: The lattice of all subsets of a 4-element set.

6.3 Fuzzy sets

The logic of power sets is based upon the Boolean algebra. Indeed, we have represented the power set 2^{Ω} as the product of multiple copies of the Boolean truth lattice – one copy for each element of Ω. We next replace this binary logic with fuzzy logic based upon the work of Zadeh [13]. All we really need to do is replace the indicator function $1_A : \Omega \to \{0,1\}$ with a “fuzzy map” $\alpha : \Omega \to [0,1]$. Each such fuzzy map represents a *fuzzy set* in Ω. Therefore, the category of all fuzzy sets is given by $[0,1]^{\Omega}$, where the unit interval $[0,1]$ is either discrete or a continuum.

The algebra of fuzzy sets is transparent. The disjunction $\alpha \vee \beta$ or “union” of two fuzzy sets α and β is defined by

$$\langle \alpha \vee \beta \rangle(x) := \max\{\alpha(x), \beta(x)\}$$

for every $x \in \Omega$; and the conjunction $\alpha \wedge \beta$ or "intersection" of two fuzzy sets is defined by

$$\langle \alpha \wedge \beta \rangle(x) := \min\{\alpha(x), \beta(x)\}$$

for every $x \in \Omega$. The involution $\sim \alpha$ of fuzzy set α is defined by

$$\langle \sim \alpha \rangle(x) := 1 - \alpha(x)$$

for every $x \in \Omega$. The category of all fuzzy sets of Ω is a distributive de Morgan algebra, but it is not a Boolean algebra. Indeed, both the laws of excluded middle and noncontradiction clearly fail.

Lemma 6.3. *Suppose Ω is the universe of all possible elements. Then the lattice order on the fuzzy sets of Ω is given by $\alpha \succcurlyeq \beta$ if and only if $\alpha(x) \geq \beta(x)$ for every $x \in \Omega$.*

This order is the same as the product order of vector lattice theory. Taking this perspective, each element $x \in \Omega$ marks a different dimension, and the fuzzy set α is a vector in $[0,1]^{\Omega}$. Thus, $\alpha(x)$ is the coordinate of α in dimension x.

In the case of discrete fuzzy logic, the codomain is understood not as the entire continuum of the unit interval, but as a discrete subset of the unit interval. Suppose the discrete fuzzy logic is based upon a total order of n distinct objects. Evoking the standard score as the truth function, the discrete unit interval is given by $\{0, (n-1)^{-1}, 2(n-1)^{-1}, \ldots, 1\}$. We conveniently represent this set as $[0,1]$ as well. Thus, the contents of $[0,1]$ can be grasped only by the context of the logical framework, e. g., discrete or continuous.

The conventional interpretation of fuzziness is that it measures vagueness. A truth-value near 1 is "vaguely true" and a truth-value near 0 is "vaguely false." Thus, a truth-value of 1/2 is the vaguest level of truth. A similar interpretation applies to a fuzzy set. In essence, a fuzzy set is a collection of fuzzy elements, and a fuzzy element is marked by the level of its fuzziness. If the fuzziness is near 1, then the fuzzy element is vaguely a member of the fuzzy set; if the fuzziness is near 0, then it is vaguely not. Rather it is a fuzzy member of the complement of the fuzzy set.

Example. We next examine how the theory of fuzzy sets works with the logic of paradox. The standard scores of the three truth-values are given by

Truth	Nonemptiness	Middle	Empty
Standard Score	1	1/2	0

This is the three-valued fuzzy logic denoted by $\mathcal{P}$. Let $\Omega = \{a_1, a_2\}$. Then the subsets of Ω are 2-vectors of numbers in the discrete unit interval $[0,1] = \{0, 1/2, 1\}$. Figure 6.4 presents the Hasse diagram of the fuzzy set lattice $\mathcal{P}^2$.

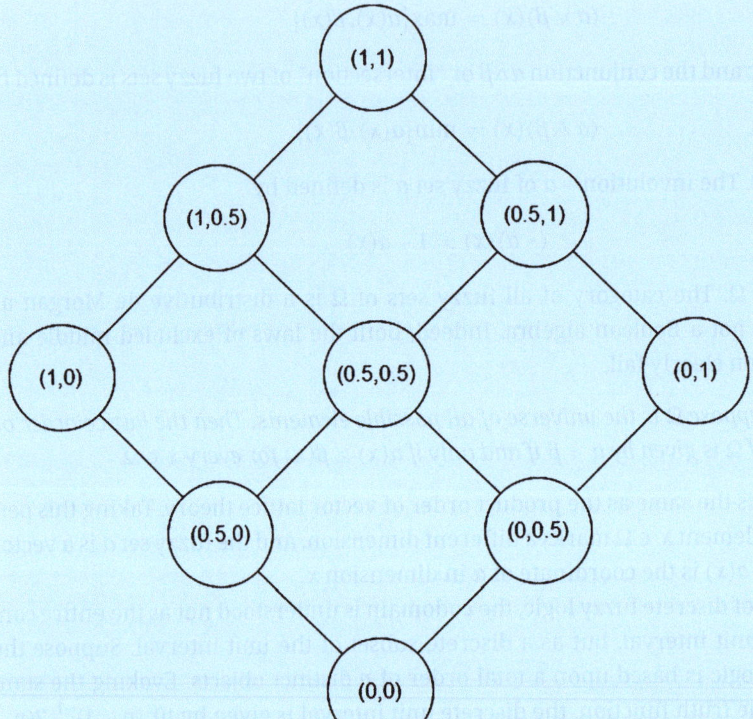

Figure 6.4: A lattice of discrete fuzzy subsets of a 2-element set.

The fuzzy set (0.5, 0.5) is the only self-negating object. There are several subalgebras of this product algebra that are isomorphic to the logic of paradox:

$$\{(1,1), (0.5,0.5), (0,0)\};$$
$$\{(1,0.5), (0.5,0.5), (0,0.5)\};$$
$$\{(0.5,1), (0.5,0.5), (0.5,0)\}.$$

All of them have the same middle. The four corners also compose a subalgebra, i. e., the tetralemma of nonfuzzy sets.

In the Buddhist logic of the T'ien-tai school, products are commonly used to convey the idea of "mutual inclusion" or "mutual possession." In the context of T'ien-tai's three-truth theory, the above Hasse diagram represents the "nine subdivisions." Since each truth includes all three truths, there are nine possible distinctions. Yet, the three truths within one truth are the same as the three truths within another truth. The product thus embodies the *unity* of the three truths. They are three-in-one and one-in-three. There is only one truth, but three ways to understand it. This is a radical departure from the interpretation of fuzzy sets in terms of vagueness.

6.4 Proofs to this chapter

Theorem 6.1. *Suppose X is a finite de Morgan algebra. Then X is isomorphic to a discrete fuzzy logic if and only if the lattice order on X is total. If the lattice order is total, then the standard score and canonical score are numerically identical measurements, and either one of them represents the discrete fuzzy logic.*

Proof. First, assume X is isomorphic to a discrete fuzzy logic. Since the fuzzy logic is total, the lattice order on X is total. Second, assume the lattice order on X is total. Appealing to Theorem 3.1, we obtain the identities

$$\Phi(x \vee y) = \max\{\Phi(x), \Phi(y)\};$$
$$\Phi(x \wedge y) = \min\{\Phi(x), \Phi(y)\}.$$

Thus, X is isomorphic to a fuzzy logic with isomorphism Φ. Appealing to Theorem 3.5, we obtain the lattice identities

$$Z(x \vee y) = \max\{Z(x), Z(y)\};$$
$$Z(x \wedge y) = \min\{Z(x), Z(y)\}.$$

Thus, X is also isomorphic to a fuzzy logic with isomorphism Z. We next verify that the isomorphisms Φ and Z are identical in this case. Since the lattice order on X is total, there is only one maximal chain from the bottom to the top. Let m denote the length of this chain. Suppose the rank of element x is i. Then $d(x) = i$ and $u(x) = m - i + 1$. Therefore,

$$\Phi(x) = \frac{i-1}{m-1} = Z(x).$$

These calculations verify that Φ and Z are identical maps. □

Lemma 6.2. *Suppose X is a finite total lattice. If* $Z(x) = r$, *where* $r \in [0,1]$, *then there exists a unique object* $y \epsilon X$ *such that* $Z(y) = 1 - r$.

Proof. Suppose the length of the maximal chain is m. Let i denote the rank of x solving $Z(x) = r$. Select y with rank $m - i + 1$. Then

$$Z(x) + Z(y) = \frac{i-1}{m-1} + \frac{m-i}{m-1} = 1.$$

Thus, we obtain $Z(y) = 1 - Z(x)$. To verify uniqueness, assume $Z(y_1) = Z(y_2)$. Then it follows that $i_1 = i_2$, where i_1 is the rank of y_1 and i_2 is the rank of y_2. Since there is only one maximal chain from the bottom $\perp$ to the top $\top$, we obtain $y_1 = y_2$. □

Lemma 6.3. *Suppose* Ω *is the universe of all possible elements. Then the lattice order on the fuzzy sets of* Ω *is given by* $\alpha \succcurlyeq \beta$ *if and only if* $\alpha(x) \geq \beta(x)$ *for every* $x \in \Omega$.

Proof. By definition, $\alpha \succcurlyeq \beta$ if and only if $\alpha = \alpha \vee \beta$, i. e., $\alpha(x) = \max\{\alpha(x), \beta(x)\}$. The latter condition holds if and only if $\alpha(x) \geq \beta(x)$ for every $x \in \Omega$. □

6.5 Exercises

(1) Construct a Hasse diagram for $\mathcal{B}^3$. Set up a table matching the subsets of the universe $\{a_1, a_2, a_3\}$ with their vector representation as 3-tuples of zeros and ones.

(2) Construct a Hasse diagram for de Morgan algebra $\mathcal{P}^2$ of fuzzy sets with the nodes labeled as vectors. Find the fixed points of the left–right automorphism. Determine all invariant subalgebras.

(3) Consider the de Morgan algebra $\mathcal{P}^3 := \mathcal{P} \times \mathcal{P} \times \mathcal{P}$ of fuzzy sets. Count the number of objects in each of the following categories: The category of nonempty-only elements, the category of empty-only elements, the category of both empty and nonempty elements, as well as the category of neither empty nor nonempty elements. Generalize your results to $\mathcal{P}^n$, where $n \geq 1$ is a natural number.

(4) Demonstrate that $(1/2, 1/2, 1/2)$ is both a canonical median and a standard median in $\mathcal{P}^3$.

(5) Consider the Boolean algebra $\mathcal{B}^n$, where n is a natural number. Explain why the number of objects of rank $k + 1$ is given by the binomial coefficient

$$\binom{n}{k} := \frac{n!}{(k)!(n-k)!},$$

i. e., the number of combinations of n distinct objects selected k at a time. Verify that the total number of objects in $\mathcal{B}^n$ is 2^n. Conclude that the following identity holds:

$$\sum_{k=0}^{n} \binom{n}{k} = 2^n.$$

(6) Define the map $f : \mathcal{P} \to \mathcal{P}$ by $f(x) = {\sim}x$. Carefully explain why this map is a dual automorphism, but not an automorphism.

(7) Construct the algebra $\mathcal{G}$ in the following manner: First, the lattice order is defined by $0 \succ 1/2 \succ 1$. Second, the involution $g : \mathcal{G} \to \mathcal{G}$ is defined by

$$g(0) = 1; \quad g(1) = 0; \quad g(1/2) = 1/2.$$

Demonstrate directly that $\mathcal{G}$ is isomorphic to $\mathcal{P}$ by constructing the appropriate isomorphism.

(8) Demonstrate that (continuous) fuzzy logic is a distributive de Morgan algebra. Utilize this result to verify that the algebra $[0, 1]^n$ of fuzzy sets is also a distributive de Morgan algebra.

(9) Verify that the diamond equation for fuzzy logic is given by

$$\max\{x, y\} + \min\{x, y\} = x + y.$$

Explain why this equation is an identity without appealing to modularity.

(10) The Hasse diagram for X^2, where X is the Belnap logical lattice, is identical to the Hasse diagram for $\mathcal{B}^4$. Explain how it is different by classifying the 16 elements into the four transcendental categories: nonempty-only, empty-only, both, and neither. Set up a table that counts the number of objects in each category for X^2 versus $\mathcal{B}^4$.

7 Down-set representations

The elegant representation theory of fuzzy logic does not generalize to the nonfuzzy case. A more general approach to the problem consists of associating the objects in a bounded lattice with subsets of a universal set so that disjunction corresponds to union and conjunction corresponds to intersection. The original breakthrough for this agenda was achieved by Birkhoff [14]. He constructed a theory that represents the objects of a finite distributive lattice with "join-irreducible" sets. The assumption of distributivity seems unavoidable for these kinds of theories, because the algebra of sets is distributive.

Although the principal down-sets come "close" to providing a complete set-theoretic representation, they do not seem to capture disjunction as well, as one might hope. Thus, we devise a slightly more general theory of finitary down-sets. The investigation of these matters provides some tools that are useful in the next chapter.

7.1 Down-sets

The representation theory that we will study is based on down- and up-sets. The down-sets suggest one kind of representation and the up-sets suggest another. These two approaches are equivalent by virtue of duality. We will examine the theory of down-sets in detail, and appeal to duality to outline a parallel theory of up-sets.

We begin with a review of the terminology. We define a *down-set* as any subset α of the bounded lattice X such that $a \in \alpha$ and $a \succcurlyeq x$ implies $x \in \alpha$. If the down-set α is also a sublattice, then it is called an *ideal*. The down-sets of the special form $D(a)$ for some $a \in \alpha$ are ideals, and they are usually called the *principal ideals*. Since we do not extensively rely upon ideal theory, we usually refer to them as the "principal down-sets" instead. However, a reader who is familiar with ideal theory will recognize many similarities. By definition, the element $a \in \alpha$ is a *maximal element* in α provided that there does not exist $x \in \alpha$ such that $x \succ a$. These extremal elements do not necessarily exist for a particular subset α, and they are not necessarily unique even if they exist. However, they always exist provided that the lattice is finite.

We next extend the basic logical operators from elements to subsets element-wise. Suppose α and β are subsets of the de Morgan algebra X. Then we define

$$\alpha \vee \beta := \{x \vee y : x \in \alpha, y \in \beta\};$$
$$\alpha \wedge \beta := \{x \wedge y : x \in \alpha, y \in \beta\};$$
$$\sim \alpha := \{\sim x : x \in \alpha\}.$$

These operators are simply the "element-wise disjunction," "element-wise conjunction," and "element-wise involution" of sets. However, there is no possibility of representing

https://doi.org/10.1515/9783112235812-007

element-wise involution as a set-theoretic complement in the general de Morgan algebra. For example, the subset $\{a, \sim a\}$ is self-negated by involution, but its complement consists of all other elements in the algebra. Hence, the following theory will focus on the lattice structure and temporarily ignore the problem of representing the involution.

Lemma 7.1. *If X is a lattice, then $D(a \wedge b) = D(a) \wedge D(b)$. If X is a distributive lattice, then $D(a \vee b) = D(a) \vee D(b)$.*

Example. The pentalemma is a simple counterexample against removing the assumption of distributivity in the second part of this lemma. Figure 7.1 below is similar to Figure 3.2.

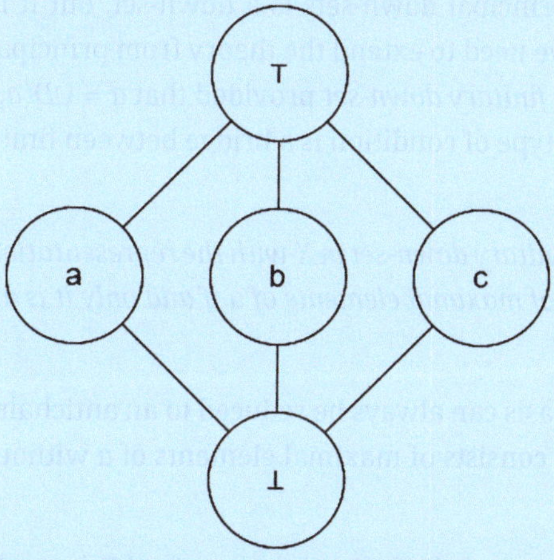

Figure 7.1: The Hasse diagram of a pentalemma.

We observe that $D(a) = \{a, \bot\}$ and $D(b) = \{b, \bot\}$. We thus compute

$$D(a) \vee D(b) = \{a \vee b, a \vee \bot, b \vee \bot, \bot \vee \bot\} = \{\top, a, b, \bot\}.$$

On the other hand, we obtain $D(a \vee b) = D(\top) = X$. We therefore conclude that $c \in D(a \vee b)$ but $c \notin D(a) \vee D(b)$.

Corollary. *If X is a distributive lattice, then the category of all principal down-sets $\{D(a) : a \in X\}$ is a lattice with respect to the element-wise lattice operators on X. The element-wise lattice order is defined by $D(a) \succcurlyeq D(b)$ if and only if $a \succcurlyeq b$.*

This result provides the traces of a representation theory for lattices. However, we would also like to transform disjunctions into unions and conjunctions into intersections, if it is feasible.

Lemma 7.2. *Suppose X is a lattice. Then $D(a) \wedge D(b) = D(a) \cap D(b)$, and $D(a) \succcurlyeq D(b)$ if and only if $D(b) \subseteq D(a)$.*

Example. Returning to the previous example, we compute that

$$D(a) \wedge D(b) = \{a \wedge b, a \wedge \bot, b \wedge \bot, \bot \wedge \bot\} = \{\bot\} = D(a) \cap D(b).$$

Furthermore, $D(a) \succcurlyeq D(\bot)$, because $D(\bot) = \{\bot\} \subseteq D(a)$.

Thus, the conjunction of two principal down-sets is equal to their intersection. This result has nothing to say about the disjunction of two principal down-sets, because there is an obstacle to completing this type of representation.

Theorem 7.3. *Suppose X is a finite lattice. Then the subset α is a down-set if and only if* $\alpha = \cup D(a_i)$, *where* $\{a_1, a_2, \ldots, a_n\}$ *are the maximal elements of* α.

In general, the union of two or more principal down-sets is a down-set, but it is not necessarily a principal down-set. Thus, we need to extend the theory from principal down-sets to all down-sets. We say that α is a *finitary down-set* provided that $\alpha = \cup D(a_i)$ for some finite *basis* $\{a_1, a_2, \ldots, a_n\} \subseteq \alpha$. This type of condition is a bridge between finite and infinite lattices.

Lemma 7.4. *Suppose X is a lattice and α is a finitary down-set in X with the representation* $\alpha = \cup D(a_i)$. *The basis* $\{a_1, a_2, \ldots, a_n\}$ *consists of maximal elements of α if and only it is an antichain, i. e.,* $a_i \asymp a_j$ *for every* $i \neq j$.

Since $a_i \succ a_j$ implies $D(a_j) \subseteq D(a_i)$, the basis can always be reduced to an antichain. Hence, we can always assume that the basis consists of maximal elements of α without loss of generality.

Lemma 7.5. *Suppose X is a lattice. The category of all (finitary) down-sets of X is a sublattice of* 2^X *endowed with its set-theoretic lattice operators of union and intersection.*

The category of all finitary down-sets is the smallest sublattice of 2^X (endowed with its set-theoretic lattice operators) containing the category of all principal down-sets. In the special case when X is finite, it coincides with the lattice of all down-sets. We next offer a partial solution to the problem of representing disjunction set-theoretically.

Lemma 7.6. *Suppose X is a lattice. Then* $D(a \vee b)$ *is the smallest ideal α in X such that* $D(a) \cup D(b) \subseteq \alpha$.

Although we cannot complete this type of representation theory, we still find the finitary down-sets to be a very useful tool. It is actually a representation theory for the antichains in X. In essence, we associate each antichain $\{a_1, a_2, \ldots, a_n\}$ with the down-set $\alpha = \cup D(a_i)$. An element thus corresponds to a singleton, which is a mathematically trivial antichain.

Lemma 7.7. *Suppose X is a bounded lattice. Then the finitary down-set α has a unique maximum if and only if α is an ideal.*

This result has enormous consequences. If we wish to study finitary down-sets with multiple maximal elements, then we should abandon ideal theory.

7.2 Up-sets

The dual theory of up-sets is next. First, we define an *up-set* as any subset β of X such that $b \in \beta$ and $x \succcurlyeq b$ implies $x \in \beta$. Filters are the duals of the ideals. If the up-set β is also a sublattice, then it is called a *filter*. In particular, the up-sets of the special form $U(b)$ for some $b \in \beta$ are filters known as the *principal filters*. We also refer to a principal filter as a "principal up-set" when we wish to emphasize its special role in the theory of up-sets. The element $b \in \beta$ is a *minimal element* in β provided that there does not exist $x \in \beta$ such that $b \succ x$.

All dual results will be prefixed "dual," but otherwise identified the same way as their dual counterparts. This duality is between disjunction $\vee$ and conjunction $\wedge$, but not between the union and intersection of sets.

Dual Lemma 7.1. *If X is a lattice, then $U(a \vee b) = U(a) \vee U(b)$. If X is a distributive lattice, then $U(a \wedge b) = U(a) \wedge U(b)$.*

Dual Corollary. *If X is a distributive lattice, then the category of all principal filters $\{U(b) : b \in X\}$ is a lattice with respect to the element-wise extensions of the lattice operators on X.*

Dual Lemma 7.2. *Suppose X is a lattice. Then $U(a) \vee U(b) = U(a) \cap U(b)$, and $U(a) \preccurlyeq U(b)$ if and only if $U(a) \supseteq U(b)$.*

Dual Theorem 7.3. *Suppose X is a finite lattice. Then the subset β is an up-set if and only if $\beta = \cup U(b_i)$, where $\{b_1, b_2, \ldots b_m\}$ are the minimal elements of β.*

We say that β is a *finitary up-set* provided that $\beta = \cup U(b_i)$ for some finite *basis* $\{b_1, b_2, \ldots, b_m\} \subseteq \beta$.

Dual Lemma 7.4. *Suppose X is a lattice and β is a finitary up-set in X with the representation $\beta = \cup D(b_i)$. The basis $\{b_1, b_2, \ldots, b_m\}$ consists of minimal elements of β if and only it is an antichain, i. e., $b_i \simeq b_j$ for every $i \neq j$.*

Dual Lemma 7.5. *Suppose X is a lattice. The category of all (finitary) up-sets of X is a sublattice of 2^X endowed with its set-theoretic lattice operators of union and intersection.*

Dual Lemma 7.6. *Suppose X is a lattice. Then $U(a \wedge b)$ is the largest filter β in X such that $\beta \subseteq U(a) \cup U(b)$.*

Dual Lemma 7.7. *Suppose X is a bounded lattice. Then the finitary up-set β has a unique minimum if and only if β is a filter.*

There are some interesting situations in which both down- and up-sets are relevant. Suppose $a \preccurlyeq b$. Then the interval from b to a was previously defined as

$$[a, b] := \{x \in X : a \preccurlyeq x \preccurlyeq b\}.$$

Notice that $[a, b] = U(a) \cap D(b)$. This simple property provides a slightly more general definition, which applies even when the end-points are incomparable. An *interval* is now redefined as the intersection of a principal up-set and a principal down-set. It is easy to verify that $[a, b] \neq \emptyset$ implies $a \preccurlyeq b$. Further notice that $U(a) = [a, \top]$ and $D(b) = [\bot, b]$. In other words, both principal up-sets and principal down-sets are special types of intervals. This fact suggests another way to combine the lattice theory of up-sets with down-sets. By definition, a subset α of L is *convex* provided that $\{a, b\} \subseteq \alpha$ implies $[a, b] \subseteq \alpha$. Since we have slightly expanded the definition of an interval, convex sets also have a somewhat wider scope. For example, an antichain is a convex set by virtue of this definition. Indeed, $a \simeq b$ implies $[a, b] \subseteq \{a, b\}$. Any subset that is convex by the original definition of an interval remains convex under the new definition.

Lemma 7.8. *Suppose X is a lattice. Every convex subset of X is the intersection of an up-set and a down-set.*

This result is essentially due to Blyth [15]. He demonstrated that a convex sublattice is the intersection of a filter and an ideal. We have modified his result by dropping the superfluous assumption that the convex set is a lattice. As a special case of this lemma, every up-set α can be written as $\alpha = \alpha \cap D(\top)$, and every down-set β can be written as $\beta = U(\bot) \cap \beta$. Therefore, they are also convex sets.

Example. Consider the standard tetralemma with a digital representation. Figure 7.2 presents its Hasse diagram.

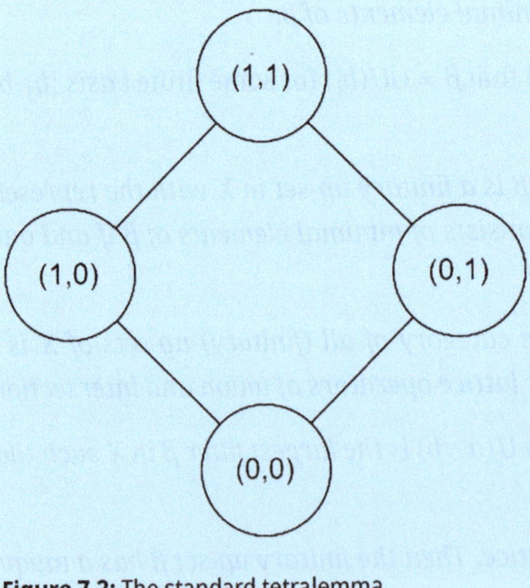

Figure 7.2: The standard tetralemma.

The middle region $\{(1,0),(0,1)\}$ is an antichain, and is therefore convex. Moreover, it is the intersection of the up-set $\{(1,1),(1,0),(0,1)\}$ and the down-set $\{(0,0),(1,0),(0,1)\}$.

Lemma 7.9. *Suppose X is a lattice. The category of all convex subsets of X is closed with respect to intersections.*

We can also construct a finitary theory of convex sets. By definition, a *finitary convex set* is the intersection of a finitary up-set and a finitary down-set. This type of set is obviously convex, but it has a finite number of minimal elements and a finite number of maximal elements. In particular, every finitary up-set and every finitary down-set is a finitary convex set. The simplest finitary convex set is an interval $[a, b]$ with $b \succ a$, which has only one minimal element $a \in X$ and one maximal element $b \in X$.

After this excursion into lattice theory, let us bring the involution back into the analysis. We have already extended the involution operator to subsets by defining it elementwise. Thus, we have $\sim \alpha := \{x \in X : \sim x \in \alpha\}$.

Theorem 7.10. *Suppose X is a de Morgan algebra. Then α is an up-set if and only if $\sim \alpha$ is a down-set. Moreover, the element a is a minimal element in α if and only if $\sim a$ is a maximal element in $\sim \alpha$; and $\alpha = \cup U(a_i)$ if and only if $\sim \alpha = \cup D(\sim a_i)$.*

This result integrates seamlessly into the theory of finitary sets. Indeed, it is obvious that the basis $\{a_1, a_2, \dots, a_n\}$ for the up-set $\alpha = \cup U(a_i)$ is an antichain if and only if the basis $\{\sim a_1, \sim a_2, \dots, \sim a_n\}$ for the down-set $\sim \alpha$ is an antichain.

7.3 Decomposition formulas

The results on down- and up-sets lead to several elegant formulas for their decomposition into principal down-sets and principal up-sets.

Lemma 7.11. *Suppose that X is a distributive lattice. If $\alpha = \cup D(a_i)$ and $\beta = \cup D(b_j)$ are finitary down-sets, then*

$$\alpha \wedge \beta = \cup_{i,j} D(a_i \wedge b_j);$$
$$\alpha \vee \beta = \cup_{i,j} D(a_i \vee b_j).$$

This result is a generalization of Lemma 7.1. A key feature of these formulas is that all combinations of i and j are required. Dual formulas hold for finitary up-sets. Thus, the next result generalizes Dual Lemma 7.1.

Dual Lemma 7.11. *Suppose that X is a distributive lattice. If $\alpha = \cup U(a_i)$ and $\beta = \cup U(b_j)$ are finitary up-sets, then*

$$\alpha \wedge \beta = \cup_{i,j} U(a_i \wedge b_j);$$
$$\alpha \vee \beta = \cup_{i,j} U(a_i \vee b_j).$$

Similar formulas are available for finitary convex sets.

Lemma 7.12. *Suppose X is a lattice. If α is a finitary convex subset with minimal elements $\{a_1, a_2, \ldots, a_n\}$ and maximal elements $\{b_1, b_2, \ldots, b_m\}$, then*

$$\alpha = \cup_{i,j}[a_i, b_j].$$

It is noteworthy that some combination of a minimal and a maximal element can be incomparable, i. e., $a_i \parallel b_j$. In this case, the interval $[a_i, b_j]$ is the empty set. This result also makes it rather obvious that the union of two convex sets is not necessarily convex. For example, the union of two intervals, say $[a_1, b_1] \cup [a_2 \cup b_2]$, excludes $[a_1, b_2] \cup [a_1 \cup b_2]$, which is typically needed to convexify the original union.

As a final lattice exercise, we devise a formula for the lattice order of sets derived from element-wise disjunction and element-wise conjunction. The following result generalizes the corollary to Lemma 7.1.

Theorem 7.13. *Suppose X is a distributive lattice. If $\alpha = \cup D(a_i)$ and $\beta = \cup D(b_j)$ are finitary down-sets, then $\alpha \succcurlyeq \beta$ if and only if $\wedge a_i \succcurlyeq \vee b_j$.*

Dual Theorem 7.13. *Suppose X is a distributive lattice. If $\alpha = \cup U(a_i)$ and $\beta = \cup U(b_j)$ are finitary up-sets, then $\alpha \succcurlyeq \beta$ if and only if $\wedge a_i \succcurlyeq \vee b_j$.*

Although this result involves only lattice theory, it meshes elegantly with an involution operator. Suppose X is a distributive de Morgan algebra. Applying the law of polarity, the condition $\wedge a_i \succcurlyeq \vee b_j$ holds if and only if $\wedge(\sim b_j) \succcurlyeq \vee(\sim a_i)$. Since up-sets are the involutions of down-sets, this relation clearly explain why the same algebraic form prevails. In the primary theorem, we are working with the maximal elements of finitary down-sets, and in the dual theorem we are working with minimal elements of finitary up-sets.

7.4 Proofs to this chapter

Lemma 7.1. *If X is a lattice, then $D(a \wedge b) = D(a) \wedge D(b)$. If X is a distributive lattice, then $D(a \vee b) = D(a) \vee D(b)$.*

Proof. Select $x \in D(a \wedge b)$, so that $x \preccurlyeq a \wedge b$. Then $x \preccurlyeq a$ implies $x \in D(a)$ and $x \preccurlyeq b$ implies $x \in D(b)$. Thus, we obtain $x = x \wedge x \in D(a) \wedge D(b)$. Conversely, select $x \in D(a)$ and $y \in D(b)$, so that $x \wedge y \in D(a) \wedge D(b)$. Then $x \preccurlyeq a$ implies $x \wedge y \preccurlyeq a$ and $y \preccurlyeq b$ implies $x \wedge y \preccurlyeq b$. Thus, $x \wedge y \preccurlyeq a \wedge b$, which yields $x \wedge y \in D(a \wedge b)$. Therefore, we arrive at $D(a \wedge b) = D(a) \wedge D(b)$.

Assume X is a distributive lattice. Select $x \in D(a \vee b)$, so that $x \preccurlyeq a \vee b$. Applying the distributive law, we obtain $x = x \wedge (a \vee b) = (x \wedge a) \vee (x \wedge b)$. Since $x \wedge a \preccurlyeq a$ and $x \wedge b \preccurlyeq b$, we obtain $x \in D(a) \vee D(b)$. Conversely, select $x \in D(a)$ and $y \in D(b)$, so that

$x \vee y \in D(a) \vee D(b)$. Then $x \preccurlyeq a \preccurlyeq a \vee b$ and $y \preccurlyeq b \preccurlyeq a \vee b$ imply $x \vee y \preccurlyeq a \vee b$. Thus, we obtain $x \vee y \in D(a \vee b)$. Therefore, we conclude that $D(a \vee b) = D(a) \vee D(b)$. □

Corollary. *If X is a distributive lattice, then the category of all principal down-sets $\{D(a) : a \in X\}$ is a lattice with respect to the element-wise lattice operators on X. The element-wise lattice order is defined by $D(a) \succcurlyeq D(b)$ if and only if $a \succcurlyeq b$.*

Proof. The first part is a restatement of Lemma 7.1. Thus, we only need to derive the lattice order. Assume $D(a) \succcurlyeq D(b)$. By the definition of the lattice order, we obtain $D(a) = D(a) \vee D(b) = D(a \vee b)$. Since we have $a \succcurlyeq a \vee b$ and $a \vee b \succcurlyeq b$, we conclude that $a \succcurlyeq b$. The converse is obvious. □

Lemma 7.2. *Suppose X is a lattice. Then $D(a) \wedge D(b) = D(a) \cap D(b)$, and $D(a) \succcurlyeq D(b)$ if and only if $D(b) \subseteq D(a)$.*

Proof. First, assume $x \in D(a) \wedge D(b)$, so that $x = c \wedge d$ for some $c \in D(a)$ and $d \in D(b)$. Then $x \preccurlyeq c$ implies $x \in D(a)$ and $x \preccurlyeq d$ implies $x \in D(b)$. Thus, we obtain $x \in D(a) \cap D(b)$. Conversely, assume $x \in D(a) \cap D(b)$. Then $x \preccurlyeq a$ and $x \preccurlyeq b$ implies $x \preccurlyeq a \wedge b$. Thus, we obtain $x \in D(a \wedge b) = D(a) \wedge D(b)$. Therefore,

$$D(a) \wedge D(b) = D(a) \cap D(b)$$

For the second part, notice that $D(a) \preccurlyeq D(b)$ if and only if $a \preccurlyeq b$ if and only if $D(a) \subseteq D(b)$. □

Theorem 7.3. *Suppose X is a finite lattice. Then the subset α is a down-set if and only if $\alpha = \cup D(a_i)$, where $\{a_1, a_2, \ldots, a_n\}$ are the maximal elements of α.*

Proof. Suppose α is a nonempty subset of L. Since α is finite, it contains maximal elements, i. e., $\{a_1, a_2, \ldots, a_n\}$ with $n \geq 1$. Assume α is a down-set. Since α always contains its maximal elements, we obtain $\alpha \supseteq \cup D(a_i)$. Let $x \in \alpha$. If element x is maximal in α, then $x = a_i$ for some i. If x is not a maximal element, then exists $y_1 \in \alpha$ such that $x \prec y_1$. Proceeding by induction, we construct a chain in α, say $y_1 \prec y_2 \prec y_3 \prec \ldots$, which must terminate in a maximal element, say a_i. Thus, we obtain $x \in D(a_i)$; and therefore, we arrive at $\alpha = \cup D(a_i)$. Conversely, $\alpha = \cup D(a_i)$ clearly implies that α is a down-set. □

Lemma 7.4. *Suppose X is a lattice and α is a finitary down-set in X with the representation $\alpha = \cup D(a_i)$. The basis $\{a_1, a_2, \ldots, a_n\}$ consists of maximal elements of α if and only it is an antichain, i. e., $a_i \asymp a_j$ for every $i \neq j$.*

Proof. Assume the basis $\{a_1, a_2, \ldots, a_n\}$ consists of maximal elements of α. Then it is not the case that $a_j \succ a_i$ or $a_i \succ a_j$, because a_i and a_j are maximal elements of α. Therefore, we conclude that $a_i \asymp a_j$ for every $i \neq j$, i. e., the basis is an antichain.

Conversely, assume the basis $\{a_1, a_2, \ldots, a_n\}$ is an antichain. Contrary to hypothesis, assume some element a_i is not maximal. Thus, there exists an element $a \in \alpha$ such that

$a \succ a_i$. Since $\alpha = \cup D(a_i)$, we must have $a \in D(a_j)$ for some element a_j. We thus obtain $a_j \succcurlyeq a \succ a_i$. However, $a_j \succ a_i$ contradicts the assumption that the basis is an antichain. ☐

Lemma 7.5. *Suppose X is a lattice. The category of all (finitary) down-sets of X is a sublattice of 2^X endowed with its set-theoretic lattice operators of union and intersection.*

Proof. We first show that the union of two down-sets and the intersection of two down-sets are down-sets. Suppose α and β are two down-sets. Assume $x \preccurlyeq y$, where $y \in \alpha \cup \beta$. Then $y \in \alpha$ or $y \in \beta$, which implies $x \in \alpha$ or $x \in \beta$. Therefore, $x \in \alpha \cup \beta$, which confirms that $\alpha \cup \beta$ is a down-set. Next assume $x \preccurlyeq y$, where $y \in \alpha \cap \beta$. Then $y \in \alpha$ and $y \in \beta$, which implies $x \in \alpha$ and $x \in \beta$. Therefore, $x \in \alpha \cap \beta$, which confirms that $\alpha \cap \beta$ is a down-set.

We next show that the union of two finitary down-sets and the intersection of two finitary down-sets are finitary down-sets. Assume $\alpha = \cup D(a_i)$ with finite basis $\{a_1, a_2, \ldots, a_n\}$, and $\beta = \cup D(b_j)$ with finite basis $\{b_1, b_2, \ldots b_m\}$. Now it is obvious that $\alpha \cup \beta$ is a finitary down-set with basis $\{a_1, a_2, \ldots, a_n, b_1, b_2, \ldots b_m\}$. Consider the intersection $\alpha \cap \beta$. The set-theoretic decomposition is given by

$$\alpha \cap \beta = \cup D(a_i) \cap D(b_j).$$

Applying Lemmas 7.1 and 7.2, we obtain $D(a_i) \cap D(b_j) = D(a_i \wedge b_j)$ for every i and j. Therefore, we conclude that $\alpha \cap \beta$ is a finitary down-set. ☐

Lemma 7.6. *Suppose X is a lattice. Then $D(a \vee b)$ is the smallest ideal α in X such that $D(a) \cup D(b) \subseteq \alpha$.*

Proof. Recall that every principal down-set is an ideal. Assume $x \in D(a) \cup D(b)$, so that $x \in D(a)$ or $x \in D(b)$. Then $x \preccurlyeq a$ or $x \preccurlyeq b$ implies $x \preccurlyeq a \vee b$. Thus, we obtain $x \in D(a \vee b)$. Therefore, the down-set $D(a \vee b)$ is an ideal such that $D(a) \cup D(b) \subseteq D(a \vee b)$. For the second part, assume α is an ideal such that $D(a) \cup D(b) \subseteq \alpha$. Then $a \in \alpha$ and $b \in \alpha$ implies $a \vee b \in \alpha$, because α is a lattice. Therefore, $D(a \vee b) \subseteq \alpha$, because α is a down-set. ☐

Lemma 7.7. *Suppose X is a bounded lattice. Then the finitary down-set α has a unique maximum if and only if α is an ideal.*

Proof. If α has a unique maximum a, then $\alpha = D(a)$, which represents it as a principal ideal. Conversely, assume α is an ideal. Since α is finitary, it has the representation $\alpha = \cup D(a_i)$ with finite basis $\{a_1, a_2, \ldots, a_n\}$. Let $a := \vee a_i$. Since α is an ideal, we obtain $a \in \alpha$. Therefore, $\alpha = D(a)$ and the element $a \in \alpha$ is its unique maximum. ☐

Lemma 7.8. *Suppose X is a lattice. Every convex subset of X is the intersection of an up-set and a down-set.*

Proof. Assume α is a convex subset of X. Consider the subsets $\upsilon = \cup_{a \in \alpha} U(a)$ and $\delta = \cup_{a \in \alpha} D(a)$. Then υ is an up-set and δ is a down-set. It is easy to verify that $\alpha = \upsilon \cap \delta$. ☐

Lemma 7.9. *Suppose X is a lattice. The category of all convex subsets of X is closed with respect to intersections.*

Proof. Let α and β denote two convex sets in X. Then $\{a, b\} \subseteq \alpha \cap \beta$ implies (1) $\{a, b\} \subseteq \alpha$ and (2) $\{a, b\} \subseteq \beta$. Condition (1) implies $[a, b] \subseteq \alpha$ and condition (2) implies $[a, b] \subseteq \beta$. Therefore, we obtain $[a, b] \subseteq \alpha \cap \beta$. □

Theorem 7.10. *Suppose X is a de Morgan algebra. Then α is an up-set if and only if $\sim \alpha$ is a down-set. Moreover, the element a is a minimal element in α if and only if $\sim a$ is a maximal element in $\sim \alpha$; and $\alpha = \cup U(a_i)$ if and only if $\sim \alpha = \cup D(\sim a_i)$.*

Proof. First, assume α is an up-set. Then $a \in \sim \alpha$ and $x \preccurlyeq a$ imply $\sim a \in \alpha$ and $\sim x \succcurlyeq \sim a$. Thus, $\sim x \in \alpha$, which implies $x \in \sim \alpha$. Therefore, $\sim \alpha$ is a down-set. The converse follows from the duality principle.

Next, observe that $a \succ x$ if and only if $\sim x \succ \sim a$, by virtue of the law of strong polarity. Therefore, a is not minimal in α if and only if $\sim a$ is not maximal in $\sim \alpha$.

The final part follows from the identity $\sim U(a_i) = D(\sim a_i)$. □

Lemma 7.11. *Suppose that X is a distributive lattice. If $\alpha = \cup D(a_i)$ and $\beta = \cup D(b_j)$ are finitary down-sets, then*

$$\alpha \wedge \beta = \cup_{i,j} D(a_i \wedge b_j);$$
$$\alpha \vee \beta = \cup_{i,j} D(a_i \vee b_j).$$

Proof. These formulas are derived computationally. First,

$$\alpha \wedge \beta = \{\cup D(a_i)\} \wedge \{\cup D(b_j)\} = \bigcup \{D(a_i) \wedge D(b_j)\} = \bigcup D(a_i \wedge b_j).$$

Second,

$$\alpha \vee \beta = \{\cup D(a_i)\} \vee \{\cup D(b_j)\} = \bigcup \{D(a_i) \vee D(b_j)\} = \bigcup D(a_i \vee b_j).$$

The last step in both of these derivations is an application of Lemma 7.1. □

Lemma 7.12. *Suppose X is a lattice. If α is a finitary convex subset with minimal elements $\{a_1, a_2, \ldots, a_n\}$ and maximal elements $\{b_1, b_2, \ldots, b_m\}$, then*

$$\alpha = \cup_{i,j} [a_i, b_j].$$

Proof. Suppose $x \in \alpha$. Since α is finite, $x \in U(a_i)$ for some minimal element $a_i \in \alpha$ and $x \in D(b_j)$ for some maximal element $b_j \in \alpha$. Thus, we obtain $x \in [a_i, b_j]$. Conversely, assume $x \in [a_i, b_j]$. Then $x \in \alpha$, because α is convex. □

Theorem 7.13. *Suppose X is a distributive lattice. If $\alpha = \cup D(a_i)$ and $\beta = \cup D(b_j)$ are finitary down-sets, then $\alpha \succcurlyeq \beta$ if and only if $\wedge a_i \succcurlyeq \vee b_j$.*

Proof. Assume $\alpha = \cup D(a_i)$ and $\beta = \cup D(b_j)$ are finitary down-sets. Without loss of generality, we may further assume their bases are antichains. Notice that $x \in \alpha$ if and only if $a_i \succcurlyeq x$ for some a_i in the basis of α. Thus, we obtain

$$a_i \vee b_j \succcurlyeq a_i \succcurlyeq x$$

for every b_j in the basis of β. Therefore, we find that $\alpha \subseteq \alpha \vee \beta$ holds universally.

By definition, $\alpha \succcurlyeq \beta$ if and only if $\alpha = \alpha \vee \beta$. In turn, this equality holds if and only if $\alpha \vee \beta \subseteq \alpha$. We next appeal to Theorem 7.10 to represent $\alpha \vee \beta$ as the union of all down-sets of the form $D(a_i \vee b_j)$. Thus, we obtain $x \in \alpha \vee \beta$ if and only if we have $x \in D(a_i \vee b_j)$ for some basis elements a_i and b_j. Similarly, we obtain $x \in \alpha$ if and only if $x \in D(a_k)$ for some basis element a_k. We must therefore verify that $a_i \vee b_j \succcurlyeq x$ implies $a_k \succcurlyeq x$. However, we only need to check this condition for $x = a_i \vee b_j$. We therefore arrive at a necessary and sufficient condition for $\alpha \vee \beta \subseteq \alpha$: For every pair a_i and b_j, there exists a_k such that $a_k \succcurlyeq a_i \vee b_j$. Further notice that $a_k \succcurlyeq a_i \vee b_j \succcurlyeq a_i$ implies $a_k = a_i$, because the basis for α was selected as an antichain. We thus obtain $a_i = a_i \vee b_j$, which means that $a_i \succcurlyeq b_j$ for every pair of basis elements a_i and b_j. This condition is equivalent to $\wedge a_i \succcurlyeq \vee b_j$. □

7.5 Exercises

(1) Suppose X is a lattice. Demonstrate that $D(a \vee b) = D(a) \cup D(b)$ if and only if $a \nparallel b$.

(2) Suppose X is a lattice. Assume $a \preccurlyeq b$ and $c \preccurlyeq d$. Compute the smallest convex set containing the intervals $[a, b]$ and $[c, d]$.

(3) Suppose X is a finite lattice. Demonstrate that the subset $\{x \in X : \Phi(x) \geq \delta\}$, where $0 \leq \delta \leq 1$, is an up-set of X.

(4) Suppose X is a de Morgan algebra. Demonstrate that the category of all self-negated objects in X is a convex set.

(5) Suppose X is a de Morgan algebra. Demonstrate that $a \in X$ is nonempty if and only if $D(a) \cap U(\sim a) \neq \emptyset$.

(6) Suppose X is a bounded lattice and $[a, b]$ is an interval in X such that

$$\top \succ b \succcurlyeq a \succ \bot.$$

Verify that the set-theoretic complement of $[a, b]$ is not a convex set.

(7) Verify that the middle region $\{(1, 0), (0, 1)\}$ of the tetralemma $\mathcal{B}^2$ is a convex set. Decompose this region into a union of intervals.

(8) Consider the de Morgan algebra $\mathcal{P}^2$ of fuzzy sets. Identify the down-set of all empty elements and up-set of all nonempty elements. Decompose these subsets into a union of (nonempty) intervals.

(9) Demonstrate that the power set 2^X is a lattice with the lattice operators defined by

$$\alpha \wedge \beta := \alpha \cap \beta,$$
$$\alpha \vee \beta := \alpha \cup \beta.$$

Furthermore, verify that 2^X is a Boolean algebra when the involution is defined by

$$\sim \alpha := \alpha'.$$

(10) Suppose X is a finite de Morgan algebra. If α is a convex subset with minimal elements $\{a_1, a_2, \ldots, a_n\}$ and maximal elements $\{b_1, b_2, \ldots, b_m\}$, demonstrate that

$$\alpha = \left\{\bigcup U(a_i)\right\} \cap \left\{\bigcup D(b_j)\right\}.$$

8 The middle region

Not everything can be neatly cataloged into the transcendental categories of emptiness and nonemptiness. The *middle region* M of the de Morgan algebra X is defined by

$$M := \{x \in X : x \simeq {\sim}x\}.$$

Thus, there are two types of objects in the middle region: self-negating objects, i. e., $x = {\sim}x$; and incomparable opposites, i. e., $x \parallel {\sim}x$. In other words, it is the combined transcendental categories of "both" and "neither." Since we associate the nonempty-only objects with the top and the empty-only objects with the bottom, we naturally associate the middle region with the "middle" of the lattice order. We are now going to investigate the middle region and explore how far this analogy takes us. In a finite de Morgan algebra, the middle region includes the medians. Our primary method of attack is to mimic the antichain of the medians and impose their algebraic properties upon the middle region as a whole.

The analysis of the middle region is much more formidable than the regions of transcendental emptiness and nonemptiness. The category of "both" presents major challenges, but the category of "neither" almost defies internal analysis. Curiously, its scope is largest in the Boolean algebra, where it consists of everything except the top and bottom; and its scope is smallest in fuzzy logic, where it is an empty category.

8.1 Convexity

Although we have rejected set-theoretic complementation as a universal representation of involution, it still has some useful purposes.

Lemma 8.1. *Suppose X is a lattice. Then the subset a is a down-set if and only its complement a' is an up-set.*

Theorem 8.2. *Suppose X is a de Morgan algebra. Both the middle region and the subset of incomparable opposites are convex sets.*

Example. The middle region of the tetralemma $\mathcal{B}^2$ is given by

$$M = \{(0,1),(1,0)\}.$$

Notice that M is the intersection of the up-set $\{(0,1),(1,0),(1,1)\}$ and the down-set $\{(0,1),(1,0),(0,0)\}$. Thus, M is a convex set.

The middle region is possibly an empty set. For example, a discrete fuzzy logic with an even number of truth-values has no values in the middle region. Hence, we seek

https://doi.org/10.1515/9783112235812-008

conditions that guarantee that the middle region is a nonempty set. There are several techniques for algebraically identifying objects in the middle region.

Theorem 8.3. *Suppose a is a minimal nonempty element of the de Morgan algebra X. If* $a \succ b \succ \sim a$, *then* $b \parallel \sim b$ *and* $a = b \vee \sim b$.

Consider the auxiliary condition $a \succ b \succ \sim a$. Since $a \succ \sim a$, the element a is nonempty-only and the element $\sim a$ is empty-only. The auxiliary condition therefore eliminates the possibility that a is self-negating. Given that another element b is strictly between them, the result builds the tetralemma displayed in Figure 8.1.

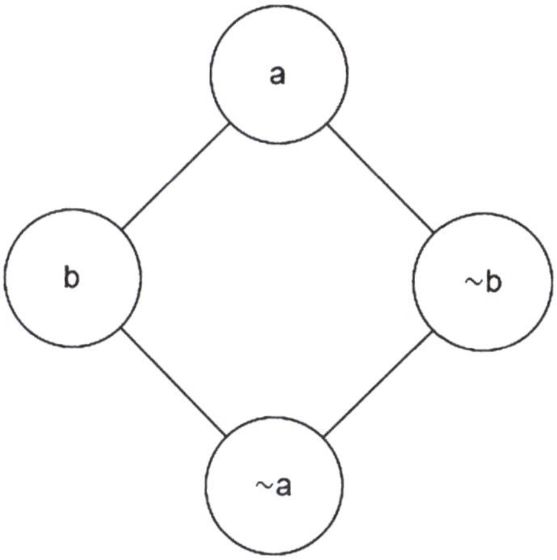

Figure 8.1: Penetration of the middle region with a tetralemma.

However, there is one qualification here. We have established by Theorem 4.3 that the category of nonempty elements is an up-set, but not necessarily a finitary up-set. Otherwise, we have provided no conditions for the existence of a minimal nonempty element. The result, as it stands, can be applied only to the finite de Morgan algebra. In this case, the category of nonempty elements is a finitary up-set.

Example. It would certainly be useful to find some type of partial converse to Theorem 8.3, but the condition $b \parallel \sim b$ does not necessarily imply that $b \vee \sim b$ is a minimal nonempty truth-value. We will construct a simple counterexample. Consider the de Morgan algebra with the Hasse diagram displayed in Figure 8.2.

The middle region consists of three elements $\{a, m, \sim a\}$. The pair of opposites $\{a, \sim a\}$ is incomparable, but their disjunction $a \vee \sim a = \top$ is not a minimal nonempty truth-value. We assume that m is self-negated, so that

$$\sim(m \vee a) = m \wedge \sim a$$

and

$$\sim(m \vee \sim a) = m \wedge a.$$

The only minimal nonempty value is thus the self-negated median m. This algebra is isomorphic to $\mathcal{P}^2$.

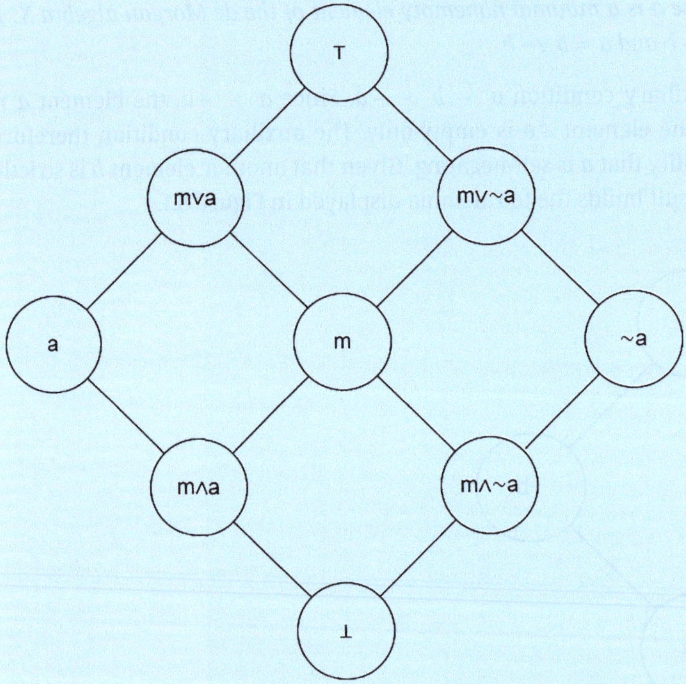

Figure 8.2: A Hasse diagram with three medians and nine nodes.

Theorem 8.4. *Suppose a and b are minimal nonempty elements in the de Morgan algebra L. If* $a \neq b$, *then* $[\sim a, a] \cap [\sim b, b] = \emptyset$.

This result is nontrivial only if $a \succ \sim a$ and $b \succ \sim b$. In this case, both a and b are nonempty-only. Of course, this restriction eliminates the Boolean algebra, which has only one nonempty element, i. e., the top $\top$. On the other hand, it guarantees that the relevant interval subalgebras exist. In this light, the theorem asserts that interval subalgebras of the form $[\sim x, x]$ for minimal nonempty elements x are pairwise disjoint.

8.2 Kleene algebra

There are several definitions of the Kleene algebra in the literature, but we have selected one that fits well into our analysis [16]. By definition, the *Kleene algebra* is a de Morgan algebra that satisfies the auxiliary condition that $x \vee \sim x \succcurlyeq y \wedge \sim y$ for every $x, y \in X$. In other words, every nonempty object is ordered above or equal to every empty object.

Although this condition seems restrictive, it is satisfied by the tetralemma, the Boolean algebra, and fuzzy logic. The next result broadens the scope of the Kleene algebra to include another type of algebra that we have studied. Moreover, the product of two Kleene algebras is another Kleene algebra. Hence, it is useful to know something about the general structure of the Kleene algebra.

Lemma 8.5. *If X is a de Morgan algebra with a negatively transitive strict lattice order, then X is a Kleene algebra.*

There is an interesting optimization principle in play here. Suppose the element c is a maximal element in the subset $C \subseteq X$. Thus, there does not exist $x \in C$ such that $x \succ c$. In terms of the codual relation, this condition translates into $c \succcurlyeq x$. In other words, element c is a *maximum*. According to Lemma 3.8, the negative transitivity of the strict order $\succ$ is equivalent to the transitivity of the codual relation $\succcurlyeq$. In this case, there can be at most one maximum relative to the codual relation $\succcurlyeq$. Therefore, the maximal element $c \in C$ is unique, i. e., it is the maximum. Likewise, any minimal element $c \in C$ is unique.

Theorem 8.6. *Suppose X is a Kleene algebra and the object $a \in X$ is self-negating. Then a is the unique self-negating object in X. Furthermore, the principal up-set $U(a)$ consists of all nonempty objects in X, and the principal down-set $D(a)$ consists of all empty objects in X.*

Example. Suppose X and Y are fuzzy logics, each with a self-negating element. Thus, they can be either continuous or discrete with an odd number of elements. Let a denote the self-negating element in X, and let b denote the self-negating element in Y. Since the factors X and Y are Kleene algebras, the product algebra $X \times Y$ is a Kleene algebra. Furthermore, the product (a, b) is the unique self-negating element in $X \times Y$. Applying Theorem 8.6, we infer that the principal up-set $U(a, b)$ consists of all nonempty elements of $X \times Y$. This result can also be derived by a different route. In fact, Lemma 5.1 asserts that (a, b) is nonempty if and only if both a and b are nonempty; and Lemma 5.4 asserts that $U(a, b) = U(a)U(b)$. Similar comments hold for the empty elements.

On the other hand, the Kleene algebra does not necessarily have a self-negating object even if the up-set of nonempty elements is a principal up-set.

Example. Consider the finite de Morgan algebra X presented in Figure 8.3.

The strict lattice order here is negatively transitive, and X is a (distributive) Kleene algebra. Here $U(c)$ is the principal up-set of nonempty elements, and $D(\sim c)$ is the principal down-set of empty elements. Notice that there is no middle region, much less any self-negating element.

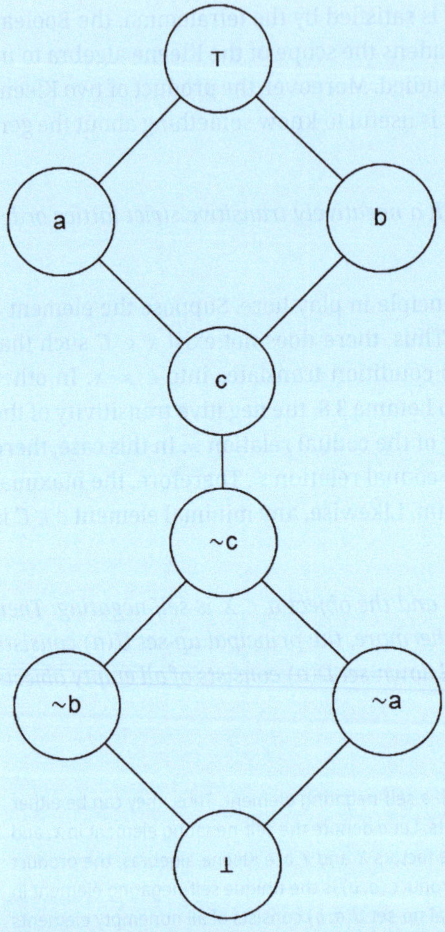

Figure 8.3: A Kleene algebra with no middle region.

The next result applies to all Kleene algebras, regardless of the existence of a self-negating element.

Theorem 8.7. *If X is a Kleene algebra, then the category of all nonempty elements is a filter and the category of all empty objects is an ideal.*

This theorem implies that the Kleene algebra cannot have more than one minimal nonempty element. Indeed, if a and b are two distinct minimal nonempty elements, then $a \wedge b$ is a nonempty element such that $a \succ a \wedge b$ and $b \succ a \wedge b$. This contradicts the assumption that a and b are minimal. Likewise, the Kleene algebra cannot have more than one maximal empty element.

Corollary. *If X is a finite Kleene algebra, then the category of all nonempty elements is a principal filter and the category of all empty objects is a principal ideal.*

Example. Consider the de Morgan algebra presented in Figure 8.4.

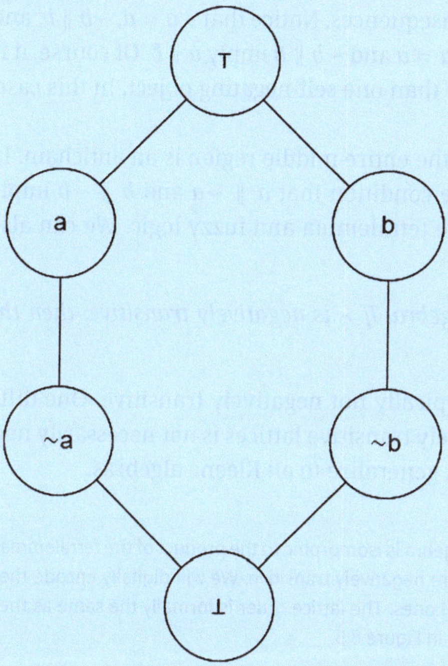

Figure 8.4: A de Morgan algebra with no middle region.

This de Morgan algebra has no middle region. The category of nonempty objects is $\{a, b, \top\}$, and the category of empty objects is $\{\sim a, \sim b, \bot\}$. This is not a Kleene algebra, because $a \parallel \sim b$ and $b \parallel \sim a$. Notice that the objects a and b are minimal nonempty elements, but $a \wedge b = \bot$. Thus, Theorem 8.6 and its corollary cannot be generalized to all de Morgan algebras.

8.3 Antichain conditions

Lemma 8.8. *Suppose X is a finite de Morgan algebra. The category of all canonical medians is an antichain.*

Corollary. *Suppose X is a finite de Morgan algebra. If a is a canonical median, then* $a \simeq \sim a$.

Thus, all canonical medians belong to the middle region, and we can think of the category of all canonical medians as the core of the middle region.

A self-negating object need not exist. For example, one does not exist in a fuzzy logic with an even number of truth-values or in a Boolean algebra. However, if one exists, it has some interesting algebraic properties. For example, if X is a finite de Morgan algebra, the self-negating condition $\sim a = a$ implies the element a is a canonical median. In general, it is always assigned to the middle region.

Lemma 8.9. *In any de Morgan algebra, $\sim a = a$ and $\sim b \simeq b$ implies $a \simeq b$.*

This lemma has some inconspicuous consequences. Notice that $\sim a = a$, $\sim b \parallel b$, and $a = b$ is an impossible combination. Thus, $\sim a = a$ and $\sim b \parallel b$ imply $a \parallel b$. Of course, it is possible that a de Morgan algebra has more than one self-negating object. In this case, $\sim a = a$, $\sim b = b$, and $a \neq b$ imply $a \parallel b$.

We next investigate the possibility that the entire middle region is an antichain. In view of Lemma 8.8, this is equivalent to the condition that $a \parallel \sim a$ and $b \parallel \sim b$ imply $a \simeq b$. This condition certainly holds for the tetralemma and fuzzy logic. We can also add another class of algebras to the list.

Lemma 8.10. *Suppose X is a de Morgan algebra. If $\succ$ is negatively transitive, then the middle region is an antichain.*

Nevertheless, the Boolean algebra is typically not negatively transitive. One difficulty here is that the product of two negatively transitive lattices is not necessarily negatively transitive. Hence, the result does not generalize to all Kleene algebras.

Example. Consider the Boolean algebra $\mathcal{B}^3$. This algebra is isomorphic to the product of the tetralemma $\mathcal{B}^2$ and the Boolean truth lattice $\mathcal{B}$, both of which are negatively transitive. We will digitally encode the elements of $\mathcal{B}^3$ as ordered three-tuples of zeros and ones. The lattice order is formally the same as the vector order. The Hasse diagram for $\mathcal{B}^3$ is presented in Figure 8.5.

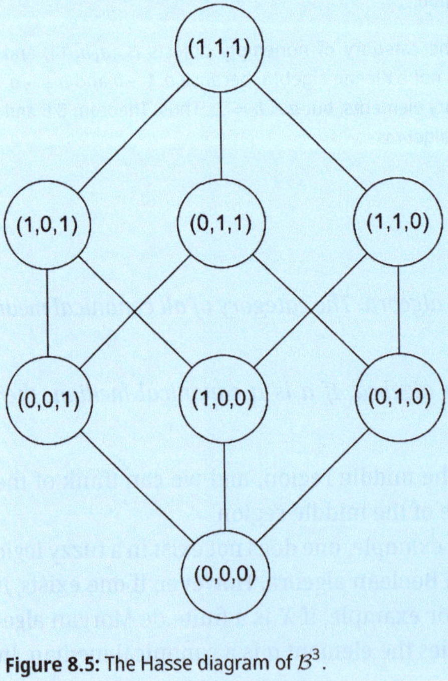

Figure 8.5: The Hasse diagram of $\mathcal{B}^3$.

For instance, notice that $(1,0,1) > (0,0,1) \parallel (0,1,0)$, but $(1,0,1) \parallel (0,1,0)$. Thus, the lattice order is not negatively transitive. In fact, $(1,0,1)$ and $(0,1,0)$ are incomparable opposites. Furthermore, the middle region, which consists of everything except the top and bottom, is not an antichain.

Theorem 8.11. *Suppose X is a de Morgan algebra. The middle region is an antichain if and only if $a \simeq {\sim} a$ and $b \succ a$ imply $b \succ {\sim} b$.*

In essence, this result asserts that the middle region is an antichain if and only if every element above one of its members is nonempty-only. This description is very intuitive. The polar law asserts that $b \succ a$ implies ${\sim} a \succ {\sim} b$. Thus, the above antecedent, i. e., $a \simeq {\sim} a$ and $b \succ a$, can be written as

$$b \succ a \simeq {\sim} a \succ {\sim} b.$$

This form reveals that the characterization in Theorem 8.11 is similar to negative transitivity, but it is somewhat weaker.

Example. It is straightforward to verify that Figure 8.6 displays a de Morgan algebra X with six elements.

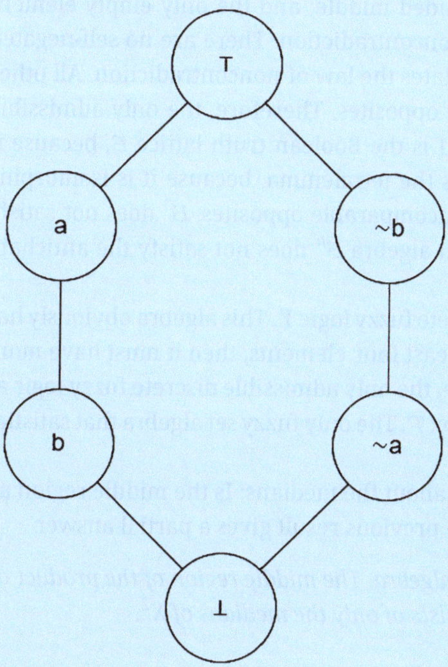

Figure 8.6: A de Morgan algebra with six elements.

The middle region $M = \{a, {\sim} a, b, {\sim} b\}$ is not antichain. Notice that X has no medians.

Curiously, the problem of characterizing the antichain condition in the product algebra is much easier to resolve in a satisfactory manner.

Theorem 8.12. *Suppose X and Y are de Morgan algebras. The middle region of the product algebra $X \times Y$ is an antichain if and only if*
(1) *Neither X nor Y has more than one nonempty-only element;*
(2) *Neither X nor Y has a pair of incomparable opposites.*

The top of the general de Morgan algebra is nonempty-only. Thus, we only need to find one other nonempty-only element in one of the factors to violate the first condition of this lemma. Except for the top and bottom, there can be no other elements outside the middle region. Regarding the second condition, once all incomparable opposites are eliminated, the only things that can remain in the middle are self-negating elements. In view of Lemma 8.8, the collection of all self-negating elements is an antichain. In summary, both X and Y must consist of a top, a bottom, and an antichain of self-negated medians. Then and only then does the product algebra satisfy the antichain condition.

For example, consider the Boolean algebra X. In general, the only nonempty element in X is the top by virtue of the law of excluded middle; and the only empty element in X is the bottom by virtue of the law of noncontradiction. There are no self-negated elements in X, because such an element violates the law of noncontradiction. All other elements in X must occur as incomparable opposites. Therefore, the only admissible Boolean algebra as a factor in Theorem 8.11 is the Boolean truth lattice $\mathcal{B}$, because it has no incomparable opposites. This covers the tetralemma, because it is isomorphic to $\mathcal{B}^2$. Since the tetralemma itself has two incomparable opposites, $\mathcal{B}^3$ does not satisfy the antichain condition. In fact, the Boolean algebra $\mathcal{B}^n$ does not satisfy the antichain condition for every $n > 2$.

For another example, consider the discrete fuzzy logic Y. This algebra obviously has no incomparable opposites. Yet, if Y has at least four elements, then it must have more than one nonempty-only element. Therefore, the only admissible discrete fuzzy logic as a factor in Theorem 8.11 is the logic of paradox $\mathcal{P}$. The only fuzzy set algebra that satisfies the antichain condition is thus $\mathcal{P}^2$.

We next return full circle to a question about the medians: Is the middle region an antichain only if it consists of medians? The previous result gives a partial answer.

Corollary. *Suppose X is a finite de Morgan algebra. The middle region of the product algebra X^2 is an antichain if and only if it consists of only the medians of X^2.*

Example. Consider the product of the logic of paradox $\mathcal{P}$, i. e., $1 > 1/2 > 0$, with the Boolean truth lattice $\mathcal{B}$, i. e., $1 > 0$. Figure 8.7 provides the Hasse diagram for the product algebra.

The middle region is given by $\{(1, 0), (0, 1)\}$. It is an antichain, but there are no medians. This example shows that the above corollary does not generalize to the case $X \times Y$ when $X \neq Y$.

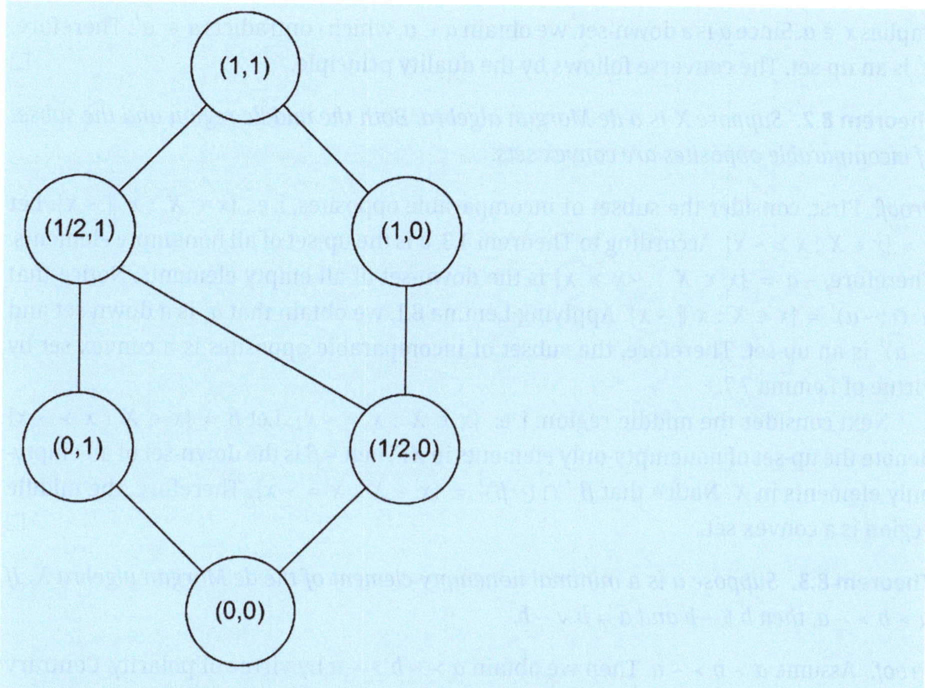

Figure 8.7: The product of the logic of paradox and Boolean truth lattice.

The following result is sometimes useful.

Lemma 8.13. *Suppose X is a finite de Morgan algebra. The following two conditions are mutually equivalent:*

(1) $a \simeq \sim a$ *implies a is a median, i. e.,* $\Phi(a) = 1/2$;

(2) $\Phi(a) > 1/2$, *implies* $a \succ \sim a$.

The first condition in this lemma asserts that every element in the middle region is a median. The second condition asserts that every element with a canonical score greater than 1/2 is nonempty-only. The converses to these two conditions always hold in a finite de Morgan algebra. However, both of these conditions fail in the previous example.

8.4 Proofs to this chapter

Lemma 8.1. *Suppose X is a lattice. Then the subset a is a down-set if and only its complement a' is an up-set.*

Proof. Assume a is a down-set. Contrary to hypothesis, assume a' is not an up-set. Then there exists $a \in a'$ and $x \succ a$ such that $x \notin a'$. Thus, set-theoretic complementation implies $x \in \alpha$. Since α is a down-set, we obtain $a \in \alpha$, which contradicts $a \in \alpha'$. Therefore, α' is an up-set. The converse follows by the duality principle. □

Theorem 8.2. *Suppose X is a de Morgan algebra. Both the middle region and the subset of incomparable opposites are convex sets.*

Proof. First, consider the subset of incomparable opposites, i. e., $\{x \in X : x \parallel {\sim}x\}$. Let $\alpha = \{x \in X : x \succcurlyeq {\sim}x\}$. According to Theorem 4.3, α is the up-set of all nonempty elements. Therefore, ${\sim}\alpha = \{x \in X : {\sim}x \succcurlyeq x\}$ is the down-set of all empty elements. Notice that $\alpha' \cap ({\sim}\alpha)' = \{x \in X : x \parallel {\sim}x\}$. Applying Lemma 8.1, we obtain that α' is a down-set and $({\sim}\alpha)'$ is an up-set. Therefore, the subset of incomparable opposites is a convex set by virtue of Lemma 7.7.

Next consider the middle region, i. e., $\{x \in X : x \simeq {\sim}x\}$. Let $\beta = \{x \in X : x \succ {\sim}x\}$ denote the up-set of nonempty-only elements in X. Then ${\sim}\beta$ is the down-set of all empty-only elements in X. Notice that $\beta' \cap ({\sim}\beta)' = \{x \in X : x \simeq {\sim}x\}$. Therefore, the middle region is a convex set. □

Theorem 8.3. *Suppose a is a minimal nonempty element of the de Morgan algebra X. If $a \succ b \succ {\sim}a$, then $b \parallel {\sim}b$ and $a = b \vee {\sim}b$.*

Proof. Assume $a \succ b \succ {\sim}a$. Then we obtain $a \succ {\sim}b \succ {\sim}a$ by virtue of polarity. Contrary to hypothesis, assume $b \nparallel {\sim}b$. Consider the case $b \succcurlyeq {\sim}b$. It follows that b is nonempty, contradicting the minimality of a. Thus, $b \parallel {\sim}b$. The case when ${\sim}b \succcurlyeq b$ is handled similarly. For the second part, notice that $a \succcurlyeq b \vee {\sim}b$. Since a is minimal, we obtain $a = b \vee {\sim}b$. □

Theorem 8.4. *Suppose a and b are minimal nonempty elements in the de Morgan algebra L. If $a \neq b$, then $[{\sim}a, a] \cap [{\sim}b, b] = \emptyset$.*

Proof. Assume $c \in [{\sim}a, a] \cap [{\sim}b, b]$. Notice that $c = a$ implies $a \in [{\sim}b, b]$. Since a and b are both minimal nonempty elements, we conclude that $a = b$ in this case. Likewise, $c = b$ implies $a = b$. Consider the case when $c = {\sim}a$. Thus, we obtain ${\sim}a \in [{\sim}b, b]$. Since ${\sim}a$ and ${\sim}b$ are both maximal empty elements, we conclude that ${\sim}a = {\sim}b$, so that $a = b$. Finally, consider the remaining case when ${\sim}a \prec c \prec a$ and ${\sim}b \prec c \prec b$. Then $a = c \vee {\sim}c$ and $b = c \vee {\sim}c$ again implies $a = b$. □

Lemma 8.5. *If X is a de Morgan algebra with a negatively transitive strict lattice order, then X is a Kleene algebra.*

Proof. Let $x = a \vee {\sim}a$ and $y = b \wedge {\sim}b$. Thus, ${\sim}x = a \wedge {\sim}a$ and ${\sim}y = b \vee {\sim}b$. We must show that $x \succcurlyeq y$. Consider the relation between x and ${\sim}y$. There are three possibilities.

Case 1: If $x \succcurlyeq {\sim}y$, then we obtain $x \succcurlyeq {\sim}y \succcurlyeq y$. Thus, we obtain $x \succcurlyeq y$.

Case 2: If ${\sim}y \succcurlyeq x$, then we obtain $x \succcurlyeq {\sim}x \succcurlyeq y$. Thus, we obtain $x \succcurlyeq y$.

Case 3: If $x \parallel {\sim}y$, then we obtain $x \parallel {\sim}y \succcurlyeq y$. Since $\succ$ is negatively transitive, we obtain $x \succcurlyeq y$ in this case. □

Theorem 8.6. *Suppose X is a Kleene algebra and the object $a \in X$ is self-negating. Then a is the unique self-negating object in X. Furthermore, the principal up-set $U(a)$ consists of all nonempty objects in X, and the principal down-set $D(a)$ consists of all empty objects in X.*

Proof. Since $a = \sim a$, element a is both empty and nonempty by virtue of Lemma 4.2. Suppose $b = \sim b$, so that b is also both empty and nonempty. We next use the defining property of the Kleene algebra. Since a is nonempty and b is empty, we obtain $a \succcurlyeq b$. Since a is empty and b is nonempty, we also obtain $b \succcurlyeq a$. Thus, we have $a = b$, which verifies the uniqueness of the self-negating object.

Next consider $U(a)$. Since X is a Kleene algebra and a is empty, x is nonempty implies $x \in U(a)$. Since a is nonempty, $x \in U(a)$ implies x is nonempty. This is an application of Theorem 4.3. Thus, $U(a)$ consists of all nonempty elements in X. Finally, consider $D(a)$. Since $a = \sim a$, we obtain $D(a) = D(\sim a) = \sim U(a)$. Thus, $D(a)$ consists of all empty elements in X. □

Theorem 8.7. *If X is a Kleene algebra, then the category of all nonempty elements is a filter and the category of all empty objects is an ideal.*

Proof. Suppose the elements a and b in X are nonempty. Thus, we have $a \succcurlyeq \sim a$ and $b \succcurlyeq \sim b$. Since $\sim a$ and $\sim b$ are empty, the defining property of the Kleene algebra yields $a \succcurlyeq \sim b$ and $b \succcurlyeq \sim a$. Thus, $a \succcurlyeq \sim a$ and $b \succcurlyeq \sim a$ imply $a \wedge b \succcurlyeq \sim a$. Likewise, $b \succcurlyeq \sim b$ and $a \succcurlyeq \sim b$ imply $a \wedge b \succcurlyeq \sim b$. Since we have $a \wedge b \succcurlyeq \sim a$ and $a \wedge b \succcurlyeq \sim b$, we obtain $a \wedge b \succcurlyeq \sim a \vee \sim b = \sim(a \wedge b)$. Therefore, $a \wedge b$ is nonempty. This verifies that the category of all nonempty elements is a filter. Since the involution of a filter is an ideal, the category of all empty objects is an ideal. □

Corollary. *If X is a finite Kleene algebra, then the category of all nonempty elements is a principal filter, and the category of all empty objects is a principal ideal.*

Proof. Since X is finite, the up-set β of all nonempty elements is finitary. Thus, it has the representation $\beta = \cup U(b_i)$, where the basis $\{b_1, b_2, \ldots, b_m\}$ is an antichain (see Dual Lemma 7.4). Let $b = \wedge_i b_i$. Then $b \in \beta$ implies $\beta = U(b)$. Therefore, β is a principal up-set. The category α of all empty sets is given by $\alpha = \sim \beta = D(\sim b)$. Therefore, α is a principal down-set. □

Lemma 8.8. *Suppose X is a finite de Morgan algebra. The category of all canonical medians is an antichain.*

Proof. Suppose a and b are canonical medians, i. e., $\Phi(a) = 1/2 = \Phi(b)$. Then the order $a \succ b$ is impossible, because it implies $\Phi(a) \succ \Phi(b)$. Likewise, $b \succ a$ is impossible. Therefore, we obtain $a \simeq b$. □

Corollary. *Suppose X is a finite de Morgan algebra. If a is a canonical median, then $a \simeq \sim a$.*

Proof. We know that $\Phi(\sim a) = 1 - \Phi(a)$ from Lemma 3.2. Thus, a is a canonical median if and only if $\sim a$ is a canonical median. □

Lemma 8.9. *In any de Morgan algebra, $\sim a = a$ and $\sim b \simeq b$ imply $a \simeq b$.*

Proof. Suppose $\sim a = a$ and $\sim b \simeq b$. Contrary to hypothesis, assume $a \succ b$. Then $\sim b \succ \sim a$ implies $\sim b \succ a \succ b$, so that $\sim b \succ b$. This contradicts $\sim b \simeq b$. The dual assumption $b \succ a$ likewise leads to a contradiction. We obtain $a \simeq b$ as the only remaining possibility. □

Lemma 8.10. *Suppose X is a de Morgan algebra. If $\succ$ is negatively transitive, then the middle region is an antichain.*

Proof. Suppose $\succ$ is negatively transitive. It follows from Lemma 3.8 that the codual relation $\succcurlyeq$ is transitive. Assume $a \simeq \sim a$ and $b \simeq \sim b$. Contrary to hypothesis, further assume $a \not\simeq b$, so that $a \succ b$ or $b \succ a$. In the case when $a \succ b$, we obtain

$$\sim b \succ \sim a \simeq a \succ b.$$

Since the codual order $\succcurlyeq$ is transitive, it follows that $\sim b \succ b$, which contradicts the initial assumption that $b \simeq \sim b$. The other case when $b \succ a$ is handled in a similar way. □

Theorem 8.11. *Suppose X is a de Morgan algebra. The middle region is an antichain if and only if $a \simeq \sim a$ and $b \succ a$ imply $b \succ \sim b$.*

Proof. First, assume the middle region M is an antichain. Further, assume $a \in M$ and $b \succ a$. If $b \in M$, then we obtain $a \simeq b$, contrary to the condition that $b \succ a$. Thus, $b \notin M$ implies b is either nonempty-only or empty-only. Since $b \succ a$, b is nonempty. Therefore, we obtain $b \succ \sim b$.

Conversely, assume $x \simeq \sim x$ and $y \succ x$ imply $y \succ \sim y$. Further, assume $a \simeq \sim a$ and $b \simeq \sim b$. Then $b \succ a$ implies $b \succ \sim b$, contradicting $b \simeq \sim b$. Therefore, it is not the case that $b \succ a$. Similarly, $a \succ b$ implies $a \succ \sim a$, contradicting $a \simeq \sim a$. Since neither $b \succ a$ nor $a \succ b$, we conclude that $a \simeq b$. □

Theorem 8.12. *Suppose X and Y are de Morgan algebras. The middle region of the product algebra $X \times Y$ is an antichain if and only if*

(1) *Neither X nor Y has more than one nonempty-only element;*
(2) *Neither X nor Y has a pair of incomparable opposites.*

Proof. Without loss of generality, let $\top$ denote the top and $\bot$ denote the bottom of X; and let $\top$ denote the top and $\bot$ denote the bottom of Y. There will be no semantic confusion, because the elements of X and Y appear as different coordinates of the product $X \times Y$. We will first establish the necessity of the two conditions separately.

(1) Assume that X has more than one nonempty-only element. One of them is the top $\top$. Let a denote the other one. Thus, we have $a \succ \sim a$. It is easy to verify that $(a, \bot) \parallel$

$(\sim a, \top)$ and $(\top, \bot) \parallel (\bot, \top)$ are two pairs of incomparable opposites such that $(\top, \bot) \succ (a, \bot)$. Therefore, the middle region of $X \times Y$ is not an antichain. A similar argument applies to Y.

(2) Assume that X has a pair of incomparable opposites, say $a \parallel \sim a$. Then we have $(a, \bot) \parallel (\sim a, \top)$ and $(\top, \bot) \parallel (\bot, \top)$ as two pairs of incomparable opposites such that $(\top, \bot) \succ (a, \bot)$. Therefore, the middle region of $X \times Y$ is not an antichain. A similar argument applies to Y.

We next demonstrate the sufficiency of the two conditions. Assume that both conditions (1) and (2) hold. This simplifies the structure of both X and Y. Let us describe it in the context of X. First, the "neither" category is an empty set. Thus, the middle region consists of only self-negated elements, and it is possibly an empty set. Moreover, the nonempty-only up-set is $\{\top\}$ and the empty-only down-set is $\{\bot\}$. In summary, there are only three possible types of elements in X: the top $\top$, the bottom $\bot$, and self-negated elements, say $m \in M$. The de Morgan algebra Y has the same generic structure. Let n denote a self-negated object in Y.

Since there are three possible types of object in each factor, there are nine possible types of products in the product algebra $X \times Y$. Some of these categories can be empty sets if one of the factors has no middle region. In this case, the factor is just a Boolean truth lattice. The top of the product is $(\top, \top)$ and the bottom of the product is $(\bot, \bot)$. This pair of opposites is comparable, i. e., $(\top, \top) \succ (\bot, \bot)$. There is also the self-negated product (m, n). The other possible types can be paired as opposites:

(i) $(\top, \bot) \parallel (\bot, \top)$;
(ii) $(m, \top) \succ (m, \bot)$;
(iii) $(\top, n) \succ (\bot, n)$.

The middle region of the product consists of three possible types of products:

$$\{(\top, \bot), (\bot, \top), (m, n)\}.$$

Notice that $(m, n) \parallel (\top, \bot)$ and $(m, n) \parallel (\bot, \top)$. Lemma 8.8 also implies that two distinct self-negated objects are incomparable. Therefore, the middle region is an antichain. □

Corollary. *Suppose X is a finite de Morgan algebra. The middle region of the product algebra X^2 is an antichain if and only if it consists of only the medians of X^2.*

Proof. If the middle region of X^2 is an antichain, then it consists of three possible types of products,

$$\{(\top, \bot), (\bot, \top), (m, m)\}.$$

Every self-negating product of type (m, m) is a canonical median by virtue of Lemma 3.7. Furthermore, we can verify that $(\top, \bot)$ and $(\bot, \top)$ are canonical medians by applying the corollary to Lemma 5.4. Thus, the middle region of X^2 contains only medians. Conversely, suppose the middle region consists of only canonical medians. Then we know it is an anti-chain from Lemma 8.8. □

Lemma 8.13. *Suppose X is a finite de Morgan algebra. The following two conditions are mutually equivalent:*

(1) $a \simeq {\sim} a$ *implies a is a canonical median, i. e.,* $\Phi(a) = 1/2$;

(2) $\Phi(a) > 1/2$, *implies* $a \succ {\sim} a$.

Proof. Assume condition (1) holds. Further, assume $\Phi(a) > 1/2$. Then it follows from condition (1) that either $a \succ {\sim} a$ or ${\sim} a \succ a$. Since a has positive valence and ${\sim} a$ has negative valence, it must be the case that $a \succ {\sim} a$. Therefore, condition (2) holds.

Conversely, assume condition (2) holds. Further, assume $a \simeq {\sim} a$. Then the case $\Phi(a) > 1/2$ implies $a \succ {\sim} a$, contradicting $a \simeq {\sim} a$. Likewise, the case $\Phi(a) < 1/2$ implies $\Phi({\sim} a) > 1/2$. Thus, ${\sim} a \succ a$, which also contradicts $a \simeq {\sim} a$. Therefore, $\Phi(a) = 1/2$, which verifies condition (1). □

8.5 Exercises

(1) Suppose X is a de Morgan algebra with middle region M. Demonstrate that an object $a \in M$ is a maximal element of M if and only if ${\sim} a$ is a minimal element of M.

(2) Suppose X is a finite de Morgan algebra and $a := \inf\{x \in X : x \succcurlyeq {\sim} x\}$. Demonstrate that X is a Kleene algebra if and only if the element a is nonempty.

(3) Suppose X is a de Morgan algebra. Verify that the middle region is an antichain if and only if $a \simeq {\sim} a$ and $b \simeq {\sim} b$ imply $a \simeq b$.

(4) Suppose X is a finite modular de Morgan algebra. Demonstrate that the middle region M is an antichain if and only if M consists of only standard medians, i. e., $M = \{x \in X : Z(x) = 1/2\}$.

(5) Suppose X is a finite modular de Morgan algebra. Demonstrate that the following two conditions are mutually equivalent:
   (i) $a \simeq {\sim} a$ implies a is a standard median, i. e., $Z(a) = 1/2$;
   (ii) $Z(a) > 1/2$, implies $a \succ {\sim} a$.

(6) Suppose X is a de Morgan algebra and $a \simeq {\sim} a$. Verify that the following four conditions are pairwise mutually equivalent:
   (i) $b \succ a$ implies $b \succ {\sim} b$;
   (ii) $b \succ {\sim} a$ implies $b \succ {\sim} b$;
   (iii) $a \succ b$ implies ${\sim} b \succ b$;
   (iv) ${\sim} a \succ b$ implies ${\sim} b \succ b$.

(7) Suppose X is a de Morgan algebra with no self-negating elements. Demonstrate that there are no subalgebras of X within the middle region M. Does this conclusion change if there is a self-negating element?

(8) Suppose X is a de Morgan algebra such that the category of all nonempty elements is a principal filter. Verify that X is a Kleene algebra.

(9) Find an example of a de Morgan algebra X with one self-negated element, say $m \in X$, such that $D(m)$ does not include all the empty elements of X. Verify that X is not a Kleene algebra.

(10) Suppose X is a fuzzy logic. Demonstrate that the algebra X^n of fuzzy sets is a Kleene algebra for every natural number $n \geq 1$.

9 Truth lattices

The idea of a "truth lattice" was introduced into logic by Gottlob Frege [17], who proposed a binary model of "true" versus "false" assigned to the propositions in a formal language. These two possibilities are called *truth-values*, but Frege took them as objects instead of opinions. Thus, truth is traditionally associated with existence and falsehood with nonexistence. This model is the simplest example of a Boolean algebra, which is reflected in the bare structure of its Hasse diagram (see Figure 9.1).

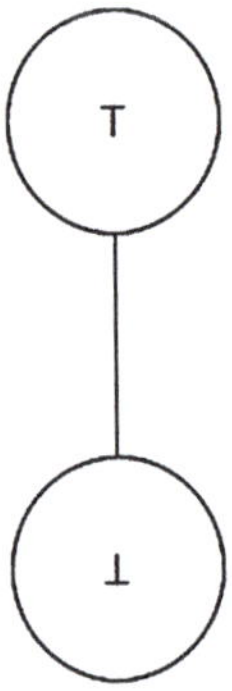

Figure 9.1: The Boolean truth lattice.

The top $\top$ represents existence and the bottom $\bot$ represents nonexistence. The involution for the Boolean truth lattice $\mathcal{B}$ is defined by $\sim\top = \bot$ and $\sim\bot = \top$.

One of the movements away from Boolean logic is to consider truth lattices with more than two truth-values. In our framework, the formal language is a grand de Morgan algebra, and the truth lattice is a smaller de Morgan algebra. There is an algebraic homomorphism from the grand algebra to the truth lattice that assigns truth-values, defined recursively from a collection of axioms. In essence, the truth lattice defines the syntax and the homomorphism defines the formal semantics.

We undertake a more modest agenda in this chapter. Our goal is to determine what kinds of algebras are associated with various truth lattices. More precisely, the general problem is to determine conditions upon the de Morgan algebra X such that there exists a homomorphism $f : X \to Y$, where Y is a fixed truth lattice. We regard this agenda as essential to justifying the truth-functional paradigm in logic. Unfortunately, there is no "universal truth lattice" that works for all de Morgan algebras. Hence, we take a piecemeal approach to the general problem. We analyze three preeminent truth lattices in depth: the Boolean truth lattice $\mathcal{B}$, the logic of paradox $\mathcal{P}$, and the tetralemma $\mathcal{B}^2$. However, we also provide some insights into a larger class of "simple truth lattices."

https://doi.org/10.1515/9783112235812-009

9.1 Homomorphisms

An essential property of a homomorphism $f : X \to Y$ is that the codomain of f, i. e., $f(X) := \{f(x) : x \in X\}$, is a subalgebra of Y. In general, a homomorphism is a many-to-one mapping. The *inverse correspondence* $f^{-1} : Y \to 2^X$ is defined by the condition that $f^{-1}(y) := \{x \in X : f(x) = y\}$. These subsets of X are called the *congruence classes* of the homomorphism f. Moreover, the binary relation on X defined by $x_1 \doteqdot x_2$ if and only if $f(x_1) = f(x_2)$ is called a *congruence relation*. It is straightforward to verify that a congruence relation $\doteqdot$ is an equivalence relation such that $a_1 \doteqdot a_2$ and $b_1 \doteqdot b_2$ implies $a_1 \vee b_1 \doteqdot a_2 \vee b_2$ and $a_1 \wedge b_1 \doteqdot a_2 \wedge b_2$.

Lemma 9.1. *If $f : X \to Y$ is a homomorphism from the lattice X to the lattice Y, then $f^{-1}(y)$ is a sublattice of X for every $y \in Y$.*

This result has an alternative formulation in the special case when X is a finite lattice, because finite lattices have tops and bottoms. In the context of Lemma 9.1, we further obtain $f^{-1}(y) = [a_y, b_y]$, where a_y and b_y are elements in X.

Lemma 9.2. *Suppose $f : X \to Y$ is a homomorphism from the bounded lattice X onto the bounded lattice Y. Then $f(\top) = \top$ and $f(\bot) = \bot$.*

This result depends critically upon the surjectivity of the homomorphism, i. e., its being onto. However, the conclusion it reaches is vital to the truth-functional paradigm of logic. In general, we say that the homomorphism $f : X \to Y$ is *truth-preserving* provided that $f(\top) = \top$. If X and Y are de Morgan algebras, then we also obtain $f(\bot) = \bot$ from the law of double negation. We next briefly examine the four transcendental categories created by the tetralemma.

Lemma 9.3. *Suppose $f : X \to Y$ is a homomorphism from de Morgan algebra X to de Morgan algebra Y. If $x \in X$ is nonempty, then $f(x) \in Y$ is nonempty; if x is empty, then $f(x)$ is empty; if x is both empty and nonempty, then $f(x)$ is both empty and nonempty.*

The category of "neither" is more problematic. Indeed, the mere fact that x is neither empty nor nonempty does not necessarily imply the same is true for its homomorphic image $f(x)$.

Example. Consider the map $f : \mathcal{B}^2 \to \mathcal{B}$ defined by

$$f(1,1) = 1;$$
$$f(1,0) = 1;$$
$$f(0,1) = 0;$$
$$f(0,0) = 0.$$

It is easy to verify that f is a homomorphism. Indeed, it is just the projection map onto the first coordinate. However, the two elements $(1,0)$ and $(0,1)$ are neither empty nor nonempty.

One criterion for designing a truth lattice Y is simplicity. We typically select one that has relatively few elements, but has a wide scope. In particular, we want one that covers a large class of de Morgan algebras with homomorphisms.

Lemma 9.4. *Suppose $f : X \to Y$ is a homomorphism from de Morgan algebra X to de Morgan algebra Y. If $x \in X$ is a self-negated element in X, then $f(x)$ is a self-negated element in Y. Furthermore, the subset $f^{-1}(y)$ is a subalgebra of X if and only if y is a self-negated element in Y.*

When we are working with self-negating elements, the logic of paradox is the simplest truth lattice possible. The discrete fuzzy logic can be regarded as a refinement of the logic of paradox. In fact, one can construct an algebraic homomorphism from a (discrete) fuzzy logic X onto the logic of paradox. Let us represent the logic of paradox $\mathcal{P}$ numerically as $1 > 1/2 > 0$ with 1 and 0 as opposites and 1/2 as self-negated. Any truth-preserving homomorphism with generic form $f : X \to \mathcal{P}$ must satisfy the basic conditions $f(\top) = 1, f(\bot) = 0$, and $f(\sim x) = \sim f(x)$ for all $x \in X$. The simplest such map assigns the value 1/2 to the fuzzy objects in X. In summary, this map $\hat{f} : X \to \mathcal{P}$ is defined by

$$\hat{f}(x) = \begin{cases} 1 & \text{if } x = \top, \\ 1/2 & \text{if } x \notin \{\bot, \top\}, \\ 0 & \text{if } x = \bot. \end{cases}$$

We next investigate the extent to which the map $\hat{f}$ is a homomorphism for the general de Morgan algebra.

Lemma 9.5. *Suppose X is a de Morgan algebra with at least three elements. The map $\hat{f} : X \to \mathcal{P}$ is a homomorphism onto the logic of paradox $\mathcal{P}$ if and only if $x \wedge y = \bot$ implies $x = \bot$ or $y = \bot$.*

The auxiliary condition in the above lemma (i. e., $x \wedge y = \bot$ implies $x = \bot$ or $y = \bot$) is equivalent to $x \vee y = \top$ implies $x = \top$ or $y = \top$. It is certainly easy to check in a fuzzy logic. In this case, we conclude that the logic of paradox is the homomorphic image of the fuzzy logic, whether its numerical range is continuous or discrete.

We next present an alternative characterization for the finite de Morgan algebra. By definition, an *atom* is the same as a successor to $\bot$, and an *antiatom* is the same as a predecessor to $\top$. Clearly, the element $x \in X$ is an atom if and only if its involution $\sim x$ is an antiatom.

Lemma 9.6. *Suppose X is a finite de Morgan algebra with at least three elements. The map $\hat{f} : X \to \mathcal{P}$ is a homomorphism onto the logic of paradox if and only if X has exactly one atom.*

Example. Consider the de Morgan algebra with the Hasse diagram displayed in Figure 9.2.

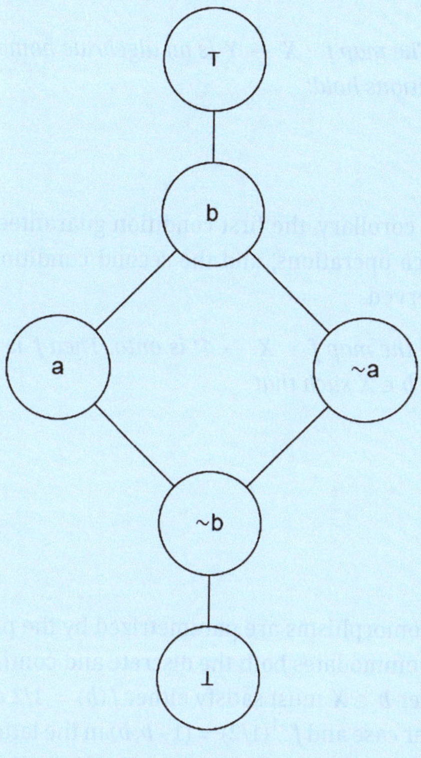

Figure 9.2: A de Morgan algebra with only one atom.

Notice that there is only one atom, i. e., $\sim b$. Therefore, the map $\hat{f}$ defines a homomorphism in this case.

These results can also be used to produce counterexamples. For example, there does not exist a surjective homomorphism $\hat{f} : \mathcal{B}^n \to \mathcal{P}$ when $n \geq 2$, because $\mathcal{B}^n$ has n atoms. In fact, there exists no homomorphism whatsoever from a Boolean algebra onto $\mathcal{P}$.

Lemma 9.7. *If X is a finite Boolean algebra, then there does not exist a homomorphism from X onto $\mathcal{P}$.*

As one might expect, the analysis can be sharpened when X is restricted to a fuzzy logic.

Lemma 9.8. *Suppose X is a fuzzy logic and Y is a de Morgan algebra. If $f : X \to Y$ is a homomorphism, then $f(X) := \{f(x) : x \in X\}$ is a fuzzy logic.*

A homomorphism $f : X \to Y$ is *isotone* or *order-preserving*, i. e., $x_1 \succcurlyeq x_2$ implies $f(x_1) \succcurlyeq f(x_2)$. In general, this condition is necessary, but not sufficient, for a lattice homomorphism. The next result exploits the totality of the fuzzy order.

Lemma 9.9. *Suppose X and Y are fuzzy logics. The map $f : X \to Y$ is a lattice homomorphism if and only if $f(x_1) > f(x_2)$ implies $x_1 \geq x_2$.*

Corollary. *Suppose X and Y are fuzzy logics. The map $f : X \to Y$ is an algebraic homomorphism if and only if the following two conditions hold:*
(1) $f(x_1) > f(x_2)$ *implies* $x_1 \geq x_2$;
(2) $f(1 - x) = 1 - f(x)$.

Breaking down the two conditions in the corollary, the first condition guarantees that the map $f : X \to Y$ preserves the lattice operations, and the second condition further guarantees that the involution is preserved.

Theorem 9.10. *Suppose X is a fuzzy logic and the map $f : X \to \mathcal{P}$ is onto. Then f is a homomorphism if and only if there exists some $b \in X$ such that*
(1) $f(b) \neq 0$;
(2) $f(1 - b) = 1 - f(b)$;
(3) $f(x) = 1$ *provided that* $1 \geq x > b$;
(4) $f(x) = 1/2$ *provided that* $b > x > 1 - b$;
(5) $f(x) = 0$ *provided that* $1 - b > x \geq 0$.

The merit of this theorem is that the homomorphisms are parametrized by the parameter $b \in X$. Furthermore, its statement accommodates both the discrete and continuous fuzzy logic. In general, the free parameter $b \in X$ must satisfy either $f(b) = 1/2$ or $f(b) = 1$. Notice $f^{-1}(1/2) = [1-b, b]$ in the former case and $f^{-1}(1/2) = (1-b, b)$ in the latter case. However, we have designed the parameter selection $b := \sup\{x \in X : f(x) = 1/2\}$ so that $f(b) = 1/2$ in the discrete case.

A natural way to construct an algebra homomorphic to $\mathcal{P}$ is to take the product algebra $X \times \mathcal{P}$, where X is any de Morgan algebra. In this case, it is obvious that the projection map $\pi_{\mathcal{F}} : X \times \mathcal{P} \to \mathcal{P}$ defined by $f(x, y) = y$ is a homomorphism.

9.2 Boolean hinges

It is well known that there exists a homomorphism $f : X \to \mathcal{B}$, where $\mathcal{B}$ is the Boolean truth lattice, provided that X is a Boolean algebra. We next investigate the possibility of generalizing this result. Suppose X is a de Morgan algebra. We say that the element $a \in X$ is a *Boolean hinge* provided that $U(a)$ and $D(\sim a)$ is a partition of X, i. e.,
(1) $U(a) \cap D(\sim a) = \emptyset$;
(2) $U(a) \cup D(\sim a) = X$.

Informally, the first condition asserts that the subsets $U(a)$ and $D(\sim a)$ are “mutually exclusive,” and the second condition asserts that they are “collectively exhaustive.” The rationale for this nomenclature is provided by the next theorem.

Theorem 9.11. *Suppose X is a de Morgan algebra. If the element $a \in X$ is a Boolean hinge, then the map $f_a : X \to \mathcal{B}$ defined by*

$$f_a(x) = \begin{cases} \top & \text{if } x \in U(a), \\ \bot & \text{if } x \in D({\sim} a) \end{cases}$$

is a homomorphism. Conversely, if X is a finite de Morgan algebra and the map $f : X \to \mathcal{B}$ is a homomorphism, then there exists a Boolean hinge $a \in X$ such that $f = f_a$.

Thus, there is great interest in unraveling the properties of a Boolean hinge. We next provide an alternative characterization in terms of coduality.

Lemma 9.12. *Suppose X is a de Morgan algebra. The following three conditions are pairwise mutually equivalent:*
(1) *Element $a \in X$ is a Boolean hinge;*
(2) *For every $x \in X$, $x \notin U(a)$ if and only if $x \in D({\sim} a)$;*
(3) *For every $x \in X$, $x \succcurlyeq a$ if and only if ${\sim} x \not\succcurlyeq a$.*

In particular, we can select $x = a$ in the third condition of this lemma, so that ${\sim} a \not\succcurlyeq a$. An empty element can therefore never be a Boolean hinge. Furthermore, $x \in U(a)$ if and only if $a \in D(x)$. Since the down-set of an empty element consists entirely of empty elements, it follows that there are no empty elements in $U(a)$.

Corollary. *Suppose X is a de Morgan algebra. If element $a \in X$ is a Boolean hinge, then the principal filter $U(a)$ contains all nonempty elements and the principal ideal $D({\sim} a)$ contains all empty elements.*

Example. Let X represent a discrete fuzzy logic with an even number of elements, say $2n$. Then the element a ranked $n + 1$ is the least nonempty element and the element ${\sim} a$ ranked n is the greatest empty element. Notice that the up-set $U(a)$ consists of all the nonempty elements of X, and the down-set $D({\sim} a)$ consists of all the empty elements of X. Since these two subsets are clearly a partition of X, the element $a \in X$ is a Boolean hinge. This example illustrates a Boolean hinge as a nonempty-only element.

Example. Let X denote a tetralemma with top $\top$, bottom $\bot$, and a pair $\{a, {\sim} a\}$ of incomparable opposites. Then $U(a) = [a, \top]$ and $D({\sim} a) = [\bot, {\sim} a]$. Since these two subsets partition X, the element a is a Boolean hinge. Likewise, the opposite element ${\sim} a$ is a Boolean hinge with $U({\sim} a) = [{\sim} a, \top]$ and $D(a) = [\bot, a]$. This example illustrates a de Morgan algebra with two Boolean hinges in the middle region.

Since a Boolean hinge cannot be an empty element, it is either nonempty or it belongs to the middle region. Suppose the Boolean hinge $a \in X$ is nonempty. Then the filter $U(a)$ consists of all nonempty elements and the ideal $D({\sim} a)$ consists of all empty elements. There is no middle region in this case.

Lemma 9.13. *Suppose X is a de Morgan algebra. If element $a \in M_X$ is a Boolean hinge, then a is a minimal element of M_X.*

We have already collected several necessary conditions for the existence of a Boolean hinge. The next result provides a partial converse.

Theorem 9.14. *Suppose X is a distributive de Morgan algebra with no self-negated objects and a is a minimal element of M_X. Then element a is a Boolean hinge if and only if $U(a)$ contains all nonempty elements of X.*

Corollary. *Suppose X is a distributive de Morgan algebra with no self-negated objects. Further, suppose the infimum b of all nonempty elements of X exists. The element $a \in M_X$ is a Boolean hinge if and only if a is a minimal element of M_X such that $b \succcurlyeq a$. In particular, every atom of a Boolean algebra is a Boolean hinge.*

The Boolean truth lattice is usually associated only with the Boolean algebra. More broadly, a homomorphism $f : X \to \mathcal{B}$, where X is a distributive de Morgan algebra, assigns a truth-value of "true" to all nonempty elements in X and a truth-value of "false" to all empty elements in X.

Example. Figure 9.3 presents a de Morgan algebra with only one Boolean hinge.

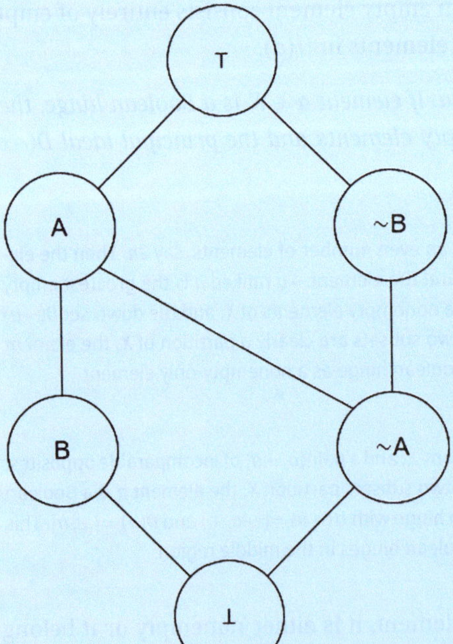

Figure 9.3: A de Morgan algebra with only one Boolean hinge.

Notice that $U(B) = \{A, B, \top\}$ and $D(\sim B) = \{\sim A, \sim B, \bot\}$ is a partition of X. Thus, element B is a Boolean hinge. In fact, it is the only Boolean hinge. The middle region is given by $M_X = \{B, \sim B\}$. Notice that the subset of all nonempty elements is given by $\{A, \top\}$. Thus, it is a non-Boolean algebra.

Lemma 9.15. *Suppose X is a finite de Morgan algebra. Let b denote the infimum of all nonempty elements of X. If b is empty, then X has no Boolean hinge.*

Example. Figure 9.4 presents the Hasse diagram of a de Morgan algebra X with no Boolean hinges.

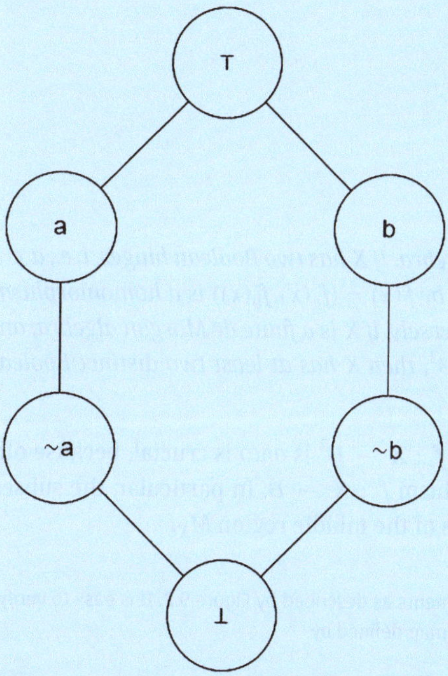

Figure 9.4: A de Morgan algebra with no Boolean hinges.

The subset of nonempty elements is given by $\{a, b, \top\}$. Its infimum is given by $\bot$. Thus, there are no Boolean hinges.

We next consider the problem of representing the truth lattice as a tetralemma. It is convenient here to represent $\mathcal{B}$ as $\{1, 0\}$ with $1 \succ 0$, and the tetralemma as $\mathcal{B}^2$. This representation emphasizes the fact that the tetralemma is a Boolean algebra. However, it does not necessarily mean that the algebra X of propositions is Boolean. We are actually interested in a homomorphism $f : X \to \mathcal{B}^2$, where X is a de Morgan algebra. Figure 9.5 is the relevant Hasse diagram for $\mathcal{B}^2$.

Notice that $\mathcal{B}^2$ has two Boolean hinges, i. e., $(1, 0)$ and $(0, 1)$. The next result is a natural extension of the previous work on the Boolean truth lattice $\mathcal{B}$. It has special significance, because $\mathcal{B}^2$ is algebraically isomorphic to the tetralemma.

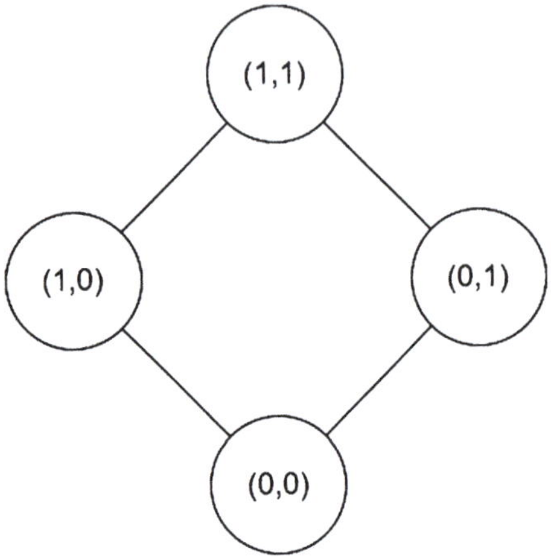

Figure 9.5: The tetralemma represented as $\mathcal{B}^2$.

Theorem 9.16. *Suppose X is a de Morgan algebra. If X has two Boolean hinges, i. e., $a \in X$ and $b \in X$, then the map $f : X \to \mathcal{B}^2$ defined by $f(x) = (f_a(x), f_b(x))$ is a homomorphism. This map is onto $\mathcal{B}^2$ if and only if $a \neq b$. Conversely, if X is a finite de Morgan algebra, and there exists a homomorphism from X onto $\mathcal{B}^2$, then X has at least two distinct Boolean hinges.*

The condition that the homomorphism $f : X \to \mathcal{B}^2$ is *onto* is crucial, because otherwise it can degenerate into a homomorphism $f : X \to \mathcal{B}$. In particular, the subsets $f^{-1}(1,0)$ and $f^{-1}(0,1)$ must be proper subsets of the middle region M_X.

Example. Consider the Kleene algebra with six elements as described by Figure 9.2. It is easy to verify that both a and $\sim a$ are Boolean hinges. The inverse map defined by

$$
\begin{aligned}
f^{-1}(1,1) &= \{b, \top\};\\
f^{-1}(1,0) &= \{a\};\\
f^{-1}(0,1) &= \{\sim a\};\\
f^{-1}(0,0) &= \{\sim b, \bot\}
\end{aligned}
$$

defines a homomorphism $f : X \to \mathcal{B}^2$.

Corollary. *Suppose X is a de Morgan algebra with two distinct Boolean hinges, say a and b. Then the homomorphism $f : X \to \mathcal{B}^2$ is well defined by*

$$
\begin{aligned}
f^{-1}(1,1) &= U(a \vee b);\\
f^{-1}(1,0) &= [a, \sim b];\\
f^{-1}(0,1) &= [b, \sim a];\\
f^{-1}(0,0) &= D(\sim a \wedge \sim b).
\end{aligned}
$$

This description of the homomorphism $f : X \to \mathcal{B}^2$ has several interesting features. First, the inverse image $f^{-1}(1,1)$ is the principal filter generated by the element $a \vee b$. Second, the inverse image $f^{-1}(0,0)$ is the principal ideal generated by the element $\sim a \wedge \sim b = \sim(a \vee b)$. Third, since the complementary intervals $[a, \sim a]$ and $[b, \sim b]$ are empty sets, the union of the other two regions (i. e., $[a, \sim b] \cup [b, \sim a]$) is a convex set.

Example. Consider the ortholattice X displayed in Figure 9.6.

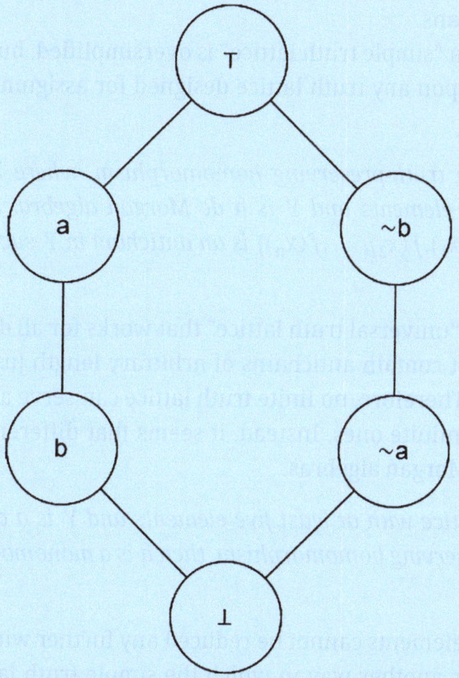

Figure 9.6: An ortholattice with two Boolean hinges.

It has two distinct Boolean hinges, i. e., b and $\sim a$. Thus, there exists a homomorphism $f : X \to \mathcal{B}^2$. Applying the corollary, we obtain

$$f^{-1}(1,1) = \{\top\};$$
$$f^{-1}(1,0) = \{a, b\};$$
$$f^{-1}(0,1) = \{\sim a, \sim b\};$$
$$f^{-1}(0,0) = \{\bot\}.$$

This example shows that the general theory covers nonmodular ortholattices.

9.3 Simple truth lattices

By definition, the de Morgan algebra X is *simple* provided that the category of nonempty-only objects is a singleton and the middle region is an antichain. If X is simple, it follows that the category of nonempty-only objects is $\{\top\}$ and the category of all empty-only objects is $\{\bot\}$. We regard these simple algebras as the best candidates for truth lattices. We always assume that a simple algebra satisfies $\top \neq \bot$ to avoid triviality. For example, the Boolean truth lattice, the logic of paradox, and the tetralemma are all "simple." These examples are simple truth lattices with zero, one, and two medians; but we intend to study those with even more than two medians.

One might think that this definition of a "simple truth lattice" is oversimplified, but the next result places severe restrictions upon any truth lattice designed for assigning truth-values in de Morgan algebras.

Theorem 9.17. *Suppose $f : X \to Y$ is a truth-preserving homomorphism, where X is a simple truth lattice with at least five elements and Y is a de Morgan algebra. If $\{x_1, x_2, \ldots, x_n\}$ is an antichain in X, then $\{f(x_1), f(x_2), \ldots, f(x_n)\}$ is an antichain in Y such that $x_i \parallel x_j$ implies $f(x_i) \parallel f(x_j)$.*

This theorem suggests that there is no "universal truth lattice" that works for all de Morgan algebras. Such a truth lattice must contain antichains of arbitrary length just to accommodate the simple truth lattices. Therefore, no finite truth lattice can serve all finite de Morgan algebras, much less all infinite ones. Instead, it seems that different truth lattices serve different classes of de Morgan algebras.

Corollary. *Suppose X is a simple truth lattice with at least five elements and Y is a de Morgan algebra. If $f : X \to Y$ is a truth-preserving homomorphism, then it is a monomorphism (i. e., it is one-to-one).*

Simple truth lattices with at least five elements cannot be reduced any further with a truth-preserving homomorphism. This is another way in which the simple truth lattices are "simple." These truth lattices are also nondistributive. Since the homomorphic image of a distributive lattice is distributive, they cannot be used as the truth lattice for a distributive de Morgan algebra.

A simple truth lattice is a prime candidate for the image of a homomorphism. It follows from Theorem 9.17 that there is no truth-preserving homomorphism from a simple truth lattice with at least five elements to a simple truth lattice with fewer elements. This result is also true to a great extent when the constraint of at least five elements is dropped. The only exception is the tetralemma $\mathcal{B}^2$, which is a simple truth lattice with four elements. The projection maps from $\mathcal{B}^2$ onto $\mathcal{B}$ are homomorphisms.

Example. The Belnap logical lattice X is also a simple truth lattice with four distinct elements. Since it has two self-negated elements, i. e., $\{T, F\}$ and $\emptyset$, there is no homomorphism onto the Boolean truth lattice

$\mathcal{B}$ by virtue of Lemma 9.4. On the other hand, assume $f : X \to \mathcal{P}$ is a truth-preserving homomorphism, where $\mathcal{P}$ is the logic of paradox. Then it must be case that $f(\{T\}) = 1, f(\{F\}) = 0$, and $f(\{T, F\}) = 1/2 = f(\emptyset)$. Notice that $f^{-1}(1/2) = \{\{T, F\}, \emptyset\}$. Since this is not a sublattice of X, we have contradicted Lemma 9.1. Therefore, there is no truth-preserving homomorphism from the Belnap logical lattice into a simple truth lattice with fewer elements.

In retrospect, the truth lattice represents an effort to formalize semantics. Nevertheless, the nuances in the construction of homomorphisms challenge the idea that "truth" can always be embodied into a truth lattice. Only the simple truth lattices with two, three, or four truth-values seem to have any utility. What can we do when the theory of truth lattices cannot be applied? In our view, the de Morgan algebra always serves as its own truth lattice.

9.4 Proofs to this chapter

Lemma 9.1. *If $f : X \to Y$ is a homomorphism from the lattice X to the lattice Y, then $f^{-1}(y)$ is a sublattice of X for every $y \in Y$.*

Proof. Assume $f : X \to Y$ is a homomorphism from the lattice X to the lattice Y. Suppose $f(x_1) = y = f(x_2)$. To verify that the lattice operations are preserved, we compute

$$f(x_1 \vee x_2) = f(x_1) \vee f(x_2) = y \vee y = y;$$
$$f(x_1 \wedge x_2) = f(x_1) \wedge f(x_2) = y \wedge y = y.$$

Thus, $f^{-1}(y)$ is a sublattice of X. □

Lemma 9.2. *Suppose $f : X \to Y$ is a homomorphism from the bounded lattice X onto the bounded lattice Y. Then $f(\top) = \top$ and $f(\bot) = \bot$.*

Proof. Since $f : X \to Y$ is a homomorphism, $x_1 \succcurlyeq x_2$ implies $f(x_1) \succcurlyeq f(x_2)$. In particular, $\top \succcurlyeq x$ implies $f(\top) \succcurlyeq f(x)$ for every $x \in X$. Since the map $f : X \to Y$ is onto, we obtain $f(\top) \succcurlyeq y$ for every $y \in Y$. Therefore, we obtain $f(\top) = \top$. A similar argument verifies that $f(\bot) = \bot$. □

Lemma 9.3. *Suppose $f : X \to Y$ is a homomorphism from de Morgan algebra X to de Morgan algebra Y. If $x \in X$ is nonempty, then $f(x) \in Y$ is nonempty; if x is empty, then $f(x)$ is empty; if x is both empty and nonempty, then $f(x)$ is both empty and nonempty.*

Proof. Assume element $x \in X$ is nonempty, i. e., $x = z \vee {\sim}z$ for some $z \in X$. Thus, we obtain $f(x) = f(z) \vee f({\sim}z) = f(z) \vee {\sim}f(z)$. Therefore, $f(x)$ is nonempty. Likewise, assume element $x \in X$ is empty, i. e., $x = z \wedge {\sim}z$ for some $z \in X$. Thus, we obtain $f(x) = f(z) \wedge f({\sim}z) = f(z) \wedge {\sim}f(z)$. Therefore, $f(x)$ is empty. Given the preservation of emptiness and nonemptiness separately, the preservation of "both" is obvious. □

Lemma 9.4. *Suppose $f : X \to Y$ is a homomorphism from de Morgan algebra X to de Morgan algebra Y. If $x \in X$ is a self-negated element in X, then $f(x)$ is a self-negated element in Y. Furthermore, the subset $f^{-1}(y)$ is a subalgebra of X if and only if y is a self-negated element in Y.*

Proof. Assume $x \in X$ is a self-negated object. Thus, we have $\sim x = x$. Since f is a homomorphism, we obtain $f(\sim x) = \sim f(x)$. But x is self-negated implies $f(\sim x) = f(x)$. Thus, we obtain $\sim f(x) = f(x)$. Therefore, $f(x)$ is a self-negated element in Y.

For the second part, assume $y \in Y$ is a self-negated element in Y. We know from Lemma 9.1 that $f^{-1}(y)$ is a sublattice of X. Furthermore, $x \in f^{-1}(y)$ implies $f(\sim x) = \sim f(x) = \sim y = y$, i. e., $\sim x \in f^{-1}(y)$. Therefore, $f^{-1}(y)$ is a subalgebra of X. Conversely, assume $f^{-1}(y)$ is a subalgebra of X. Consider any element $x \in f^{-1}(y)$. Then we obtain $y = f(\sim x) = \sim f(x) = \sim y$. Therefore, the element $y \in Y$ is self-negated. □

Lemma 9.5. *Suppose X is a de Morgan algebra with at least three elements. The map $\hat{f} : X \to \mathcal{P}$ is a homomorphism onto the logic of paradox $\mathcal{P}$ if and only if $x \wedge y = \bot$ implies $x = \bot$ or $y = \bot$.*

Proof. Assume that $x \wedge y = \bot$, but $x \neq \bot$ and $y \neq \bot$. Then we obtain $\hat{f}(x \wedge y) = 0$, but $\min\{\hat{f}(x), \hat{f}(y)\} > 0$. Therefore, $\hat{f}$ is not a homomorphism in this situation. Conversely, assume that $x \wedge y = \bot$ implies $x = \bot$ or $y = \bot$. We next verify that

$$\hat{f}(x \wedge y) = \min\{\hat{f}(x), \hat{f}(y)\}$$

by direct computation in three different cases:

Case 1. Suppose that $x \wedge y = \bot$. Without loss of generality, we take $x = \bot$, because the case $y = \bot$ is similar. We thus obtain $\hat{f}(x \wedge y) = 0$ and $\min\{\hat{f}(x), \hat{f}(y)\} = 0$.

Case 2. Suppose that $x \wedge y = \top$, so that $x = \top$ and $y = \top$. We thus obtain $\hat{f}(x \wedge y) = 1$ and $\min\{\hat{f}(x), \hat{f}(y)\} = 1$.

Case 3. Suppose that $\top \succ x \wedge y \succ \bot$, so that $\top \succ x \succ \bot$ and $\top \succ y \succ \bot$. Then $\hat{f}(x \wedge y) = 1/2$ and $\min\{\hat{f}(x), \hat{f}(y)\} = 1/2$.

In summary, $\hat{f}$ preserves conjunction. The dual condition that $x \vee y = \top$ implies $x = \top$ or $y = \top$ follows directly from the primary condition that $x \wedge y = \bot$ implies $x = \bot$ or $y = \bot$. Hence, we compute $\hat{f}(x \vee y)$ case by case as before:

Case 1. Suppose $A \vee B = \top$. Then the preceding dual condition implies $x = \top$ or $y = \top$. We thus obtain $\hat{f}(x \vee y) = 1$ and $\max\{\hat{f}(x), \hat{f}(y)\} = 1$.

Case 2. Suppose that $x \vee y = \bot$. Then $x = \bot$ and $y = \bot$. We therefore obtain $\hat{f}(x \vee y) = 0$ and $\max\{\hat{f}(x), \hat{f}(y)\} = 0$.

Case 3. Suppose that $\top \succ x \vee y \succ \bot$, so that $\top \succ x \succ \bot$ and $\top \succ y \succ \bot$. We therefore obtain $\hat{f}(x \vee y) = 1/2$, and $\max\{\hat{f}(x), \hat{f}(y)\} = 1/2$.

In summary, $\hat{f}$ preserves disjunction. Finally, it is straightforward to verify that $\hat{f}(\sim x) = 1 - \hat{f}(x)$ case by case, so that $\hat{f}$ preserves the involution operator. The map $\hat{f}$ is thus a homomorphism. □

Lemma 9.6. *Suppose X is a finite de Morgan algebra with at least three elements. The map $\hat{f} : X \to \mathcal{P}$ is a homomorphism onto the logic of paradox $\mathcal{P}$ if and only if X has exactly one atom.*

Proof. Since X is finite, it has at least one atom. We proceed by proving that X has exactly one atom if and only the condition $x \wedge y = \bot$ implies $x = \bot$ or $y = \bot$ holds. First, assume that X has exactly one atom. Suppose $x \wedge y = \bot$. Contrary to hypothesis, suppose $x \succ \bot$ and $y \succ \bot$. Since $x \succ \bot$, x dominates an atom, say $x \succcurlyeq \hat{x} \succ \bot$. Since $y \succ \bot$; y likewise dominates an atom, say $y \succcurlyeq \hat{y} \succ \bot$. Since there exists only one atom, we obtain $\hat{x} = \hat{y}$. Since $y \succcurlyeq \hat{x}$, we obtain $x \wedge y \succcurlyeq \hat{x}$. We thus arrive at the contradiction that $x \wedge y \succ \bot$. Therefore, either $x = \bot$ or $y = \bot$. This verifies that $x \wedge y = \bot$ implies $x = \bot$ or $y = \bot$.

Conversely, assume that $x \wedge y = \bot$ implies $x = \bot$ or $y = \bot$. Contrary to hypothesis, suppose L contains two distinct atoms, say $\hat{x}$ and $\hat{y}$. Then we arrive at the contradiction that $\hat{x} \succ \bot$, $\hat{y} \succ \bot$, and $\hat{x} \wedge \hat{y} = \bot$. Lemma 9.6 immediately follows from Lemma 9.5. □

Lemma 9.7. *If X is a finite Boolean algebra, then there does not exist a homomorphism from X onto $\mathcal{P}$.*

Proof. Assume that X is a finite Boolean algebra and that $f : X \to \mathcal{P}$ is a surjective homomorphism. According to Lemma 9.4, $f^{-1}(1/2)$ is a subalgebra of X, say $f^{-1}(1/2) = [\sim a, a]$. Since X is a Boolean algebra, we have $a \vee \sim a = \top$ and $a \wedge \sim a = \bot$. But $f^{-1}(1/2) \neq \emptyset$ also implies $a \succcurlyeq \sim a$. Thus, we obtain $a = \top$ and $\sim a = \bot$. Therefore, we obtain $f^{-1}(1/2) = X$, which contradicts the surjectivity of f. □

Lemma 9.8. *Suppose X is a fuzzy logic and Y is a de Morgan algebra. If $f : X \to Y$ is a homomorphism, then $f(X) := \{f(x) : x \in X\}$ is a fuzzy logic.*

Proof. Since X is fuzzy, the lattice order on X is total, i. e., $x_1 \succcurlyeq x_2$ or $x_2 \succcurlyeq x_1$. Since f is isotone, we obtain $f(x_1) \succcurlyeq f(x_2)$ or $f(x_2) \succcurlyeq f(x_1)$. Thus, the lattice order on $f(X)$ is total, i. e., $f(X)$ is fuzzy. □

Lemma 9.9. *Suppose X and Y are fuzzy logics. The map $f : X \to Y$ is a lattice homomorphism if and only if $f(x_1) > f(x_2)$ implies $x_1 \geq x_2$.*

Proof. Assume $f(x_1) \succ f(x_2)$ implies $x_1 \succcurlyeq x_2$. We need to show that the lattice operators are preserved. Since the fuzzy order on Y is complete, either $f(x_1) \succcurlyeq f(x_2)$ or $f(x_2) \succcurlyeq f(x_1)$. Without loss of generality, assume $f(x_1) \succcurlyeq f(x_2)$, so that either $f(x_1) = f(x_2)$ or $f(x_1) \succ f(x_2)$. Consider the case when $f(x_1) = f(x_2)$. Since the fuzzy order on X is complete, we have either $x_1 \succcurlyeq x_2$ or $x_2 \succcurlyeq x_1$. Assuming $x_1 \succcurlyeq x_2$, we obtain $x_1 = x_1 \vee x_2$, so that $f(x_1 \vee x_2) = f(x_1) = f(x_1) \vee f(x_2)$. Likewise, $x_2 = x_1 \wedge x_2$, so that $f(x_1 \wedge x_2) = f(x_2) = f(x_1) \wedge f(x_2)$. The same results follow when $x_2 \succcurlyeq x_1$ by virtue of a symmetric argument.

Next, consider the case when $f(x_1) \succ f(x_2)$, so that $f(x_1) = f(x_1) \vee f(x_2)$. The principal hypothesis further implies $x_1 \succcurlyeq x_2$. Thus, we obtain $x_1 = x_1 \vee x_2$, which implies $f(x_1) = f(x_1 \vee x_2)$. The conclusion is

$$f(x_1 \vee x_2) = f(x_1) = f(x_2) \vee f(x_2).$$

A similar argument shows that

$$f(x_1 \wedge x_2) = f(x_2) = f(x_1) \wedge f(x_2)$$

Therefore, the map $f : X \to Y$ is a lattice homomorphism.

Conversely, assume that $f : X \to Y$ is a lattice homomorphism. We want to show that $f(x_1) \succ f(x_2)$ implies $x_1 \succcurlyeq x_2$. Assume $f(x_1) \succ f(x_2)$. Contrary to hypothesis, further assume $x_1 \not\succcurlyeq x_2$. Since the fuzzy order is complete, we obtain $x_2 \succ x_1$. Since f is a lattice homomorphism, we arrive at $f(x_2) \succcurlyeq f(x_1)$, contradicting the assumption that $f(x_1) \succ f(x_2)$. □

Corollary. *Suppose X and Y are fuzzy logics. The map $f : X \to Y$ is an algebraic homomorphism if and only if the following two conditions hold:*

(1) $f(x_1) > f(x_2)$ *implies* $x_1 \geq x_2$*;*

(2) $f(1 - x) = 1 - f(x)$.

Proof. Condition (1) is necessary and sufficient for a lattice homomorphism. Condition (2) is necessary and sufficient for the preservation of the involution. □

Theorem 9.10. *Suppose X is a fuzzy logic and the map $f : X \to \mathcal{F}$ is onto. Then f is a homomorphism if and only if there exists some $b \in X$ such that*

(1) $f(b) \neq 0$*;*

(2) $f(1 - b) = 1 - f(b)$*;*

(3) $f(x) = 1$ *provided that* $1 \geq x > b$*;*

(4) $f(x) = 1/2$ *provided that* $b > x > 1 - b$*;*

(5) $f(x) = 0$ *provided that* $1 - b > x \geq 0$.

Proof. Assume $f : X \to \mathcal{F}$ is a homomorphism. Define $b := \sup\{x \in X : f(x) = 1/2\}$. In the discrete case, we have $f(b) = 1/2$; but in the continuous case, the relation $f(b) = 1$ is also possible. In either case, $b \in X$ and $b \geq 1/2$. We next verify the five properties:

(1) Since f is onto, there exists $x \in X$ such that $f(x) = 1/2$. If $x \leq 1/2$, then $1 - x \geq 1/2$ and $f(1 - x) = 1 - f(x) = 1/2$. Thus, we obtain $b \geq 1/2$.

(2) Since the homomorphism preserves involution, we have $f(1 - x) = 1 - f(x)$. In particular, $f(1 - b) = 1 - f(b)$.

(3) Assume $1 \geq x > b$. Contrary to hypothesis, assume $f(x) < 1$. Since $f(b) \geq 1/2$, we obtain $f(x) = 1/2$ from the monotonicity of f. However, the definition of b as a supremum implies $b \geq x$, contradicting the selection of $x > b$. Thus, we obtain $f(x) = 1$.

(4) Assume $b > x > 1 - b$. First, let $b > x \geq 1/2$. Then there exists $z \in X$ such that $b \geq z > x$ and $f(z) = 1/2$. Thus, we obtain $f(b) \geq 1/2 \geq f(x)$. Taking the involution, we obtain $f(1 - x) \geq 1/2 \geq f(1 - b)$. Therefore, we find that

$$1/2 \geq f(x) \geq f(1-x) \geq 1/2,$$

which implies $f(x) = 1/2 = f(1-x)$. Since $1/2 \geq x > 1-b$ implies $b > 1-x \geq 1/2, f(x) = 1/2$ provided that $b > x > 1-b$.

(5) Assume $1-b > x \geq 0$. Taking the involution, we obtain $1 \geq 1-x > b$. Applying case (3), we obtain $f(1-x) = 1$. Thus, we arrive at $f(x) = 0$.

Conversely, assume there exists some $b \in X$ such that the five properties are satisfied. We want to show that $f : X \to \mathcal{F}$ is a homomorphism. Since condition specifies $f(b) \neq 0$, the only two possibilities are $f(b) = 1/2$ and $f(b) = 1$. We will consider these two scenarios separately:

Scenario (A). Suppose $f(b) = 1/2$. It follows from property (2) that $f(1-b) = 1/2$. Evoking the remaining five properties, the map $f : X \to \mathcal{F}$ has the following structure:

$$f(x) = \begin{cases} 1 & \text{if } 1 \geq x > b, \\ 1/2 & \text{if } b \geq x \geq 1-b, \\ 0 & \text{if } 1-b > x \geq 0. \end{cases}$$

Notice that the third region is the involution of the first region, and the middle region is its own involution. The identity $f(1-x) = 1 - f(x)$ can thus be verified for every $x \in X$. In short, the map f preserves the involution. The criterion in Lemma 9.8 shows that f preserves the lattice operators. Therefore, the map $f : X \to \mathcal{F}$ is a homomorphism.

Scenario (B). Suppose $f(b) = 1$. It follows from property (2) that $f(1-b) = 0$. Evoking the remaining five properties, the map $f : X \to \mathcal{F}$ has the following structure:

$$f(x) = \begin{cases} 1 & \text{if } 1 \geq x \geq b, \\ 1/2 & \text{if } b > x > 1-b, \\ 0 & \text{if } 1-b \geq x \geq 0. \end{cases}$$

The verification that $f : X \to \mathcal{F}$ is a homomorphism proceeds along the same lines as Scenario (A). □

Theorem 9.11. *Suppose X is a de Morgan algebra. If the element $a \in X$ is a Boolean hinge, then the map $f_a : X \to \mathcal{B}$ defined by*

$$f_a(x) = \begin{cases} \top & \textit{if } x \in U(a), \\ \bot & \textit{if } x \in D(\sim a) \end{cases}$$

is a homomorphism. Conversely, if X is a finite de Morgan algebra and the map $f : X \to \mathcal{B}$ is a homomorphism, then there exists a Boolean hinge $a \in X$ such that $f = f_a$.

Proof. Assume element $a \in X$ is a Boolean hinge. Then the map $f_a : X \to \mathcal{B}$ is clearly well defined. Moreover, $x \in U(a)$ if and only if $\sim x \in D(\sim a)$. Thus, $f(x) = \top$ if and only if $f(\sim x) = \bot$. Therefore, we obtain $f(\sim x) = \sim f(x)$. We next verify that the lattice operations are preserved. We break this step down to three cases:

Case 1. Suppose $x_1 \in U(a)$ and $x_2 \in D(\sim a)$. First, notice that $x_1 \vee x_2 \succcurlyeq x_1$ implies $x_1 \vee x_2 \in U(a)$. Thus, $f(x_1 \vee x_2) = \top$ and $f(x_1) \vee f(x_2) = \top \vee \bot = \top$. Therefore, we obtain $f(x_1 \vee x_2) = f(x_1) \vee f(x_2)$. Second, notice that $x_2 \succcurlyeq x_1 \wedge x_2$ implies $x_1 \wedge x_2 \in D(\sim a)$. Thus, $f(x_1 \wedge x_2) = \bot$ and $f(x_1) \wedge f(x_2) = \top \wedge \bot = \bot$. Therefore, we obtain $f(x_1 \wedge x_2) = f(x_1) \wedge f(x_2)$.

Case 2. Suppose $x_1 \in U(a)$ and $x_2 \in U(a)$. Since $U(a)$ is a lattice, we further obtain $x_1 \vee x_2 \in U(a)$ and $x_1 \wedge x_2 \in U(a)$. Therefore, we compute

$$f(x_1 \vee x_2) = \top = f(x_1) \vee f(x_2),$$
$$f(x_1 \wedge x_2) = \top = f(x_1) \wedge f(x_2).$$

Case 3. Suppose $x_1 \in D(\sim a)$ and $x_2 \in D(\sim a)$. Since $D(\sim a)$ is a lattice, we further obtain $x_1 \vee x_2 \in D(\sim a)$ and $x_1 \wedge x_2 \in D(\sim a)$. Therefore, we compute

$$f(x_1 \vee x_2) = \bot = f(x_1) \vee f(x_2),$$
$$f(x_1 \wedge x_2) = \bot = f(x_1) \wedge f(x_2).$$

Since these three cases eliminate all possibilities, we conclude that $f_a : X \to \mathcal{B}$ is a homomorphism.

Conversely, assume X is finite and $f : X \to \mathcal{B}$ is a homomorphism. Since $f^{-1}(\top)$ is a finite sublattice, it has an infimum $a \in X$. Its supremum is $\top \in X$ by virtue of Lemma 9.2. Thus, we obtain $f^{-1}(\top) = [a, \top] = U(a)$. It immediately follows that $f^{-1}(\bot) = \sim f^{-1}(\top) = [\bot, \sim a] = D(\sim a)$. Since $f^{-1}(\top)$ and $f^{-1}(\bot)$ make a partition of X, element $a \in X$ is a Boolean hinge. □

Lemma 9.12. *Suppose X is a de Morgan algebra. The following three conditions are pairwise mutually equivalent:*
1. *Element $a \in X$ is a Boolean hinge;*
2. *For every $x \in X$, $x \notin U(a)$ if and only if $x \in D(\sim a)$;*
3. *For every $x \in X$, $x \succcurlyeq a$ if and only if $\sim x \not\succcurlyeq a$.*

Proof. Condition (2) is an alternative expression of the property that $U(a)$ and $D(\sim a)$ make a partition of X. Condition (3) is derived from Condition (2) by observing that $x \in D(\sim a)$ if and only if $\sim x \succcurlyeq a$. □

Corollary. *Suppose X is a de Morgan algebra. If an element $a \in X$ is a Boolean hinge, then the principal filter $U(a)$ contains all nonempty elements, and the principal ideal $D(\sim a)$ contains all empty elements.*

Proof. Assume an element $a \in X$ is a Boolean hinge. Contrary to hypothesis, assume an element $x \in U(a)$ is empty. Then $a \in D(x)$ implies a is empty. Thus, $\sim a \succcurlyeq a$ implies $a \in D(\sim a)$. However, this contradicts the definition of a Boolean hinge. Thus, x is empty implies $x \in D(\sim a)$. On the other hand, an element x being nonempty implies $\sim x$ is empty. Thus, $\sim x \in D(\sim a)$ implies $x \in U(a)$. □

Lemma 9.13. *Suppose X is a de Morgan algebra. If an element $a \in M_X$ is a Boolean hinge, then a is a minimal element of M_X.*

Proof. Assume $a \in M_X$ is a Boolean hinge. Contrary to hypothesis, assume a is not a minimal element of M_X. Thus, there exists $x \in M_X$ such that $a \succ x$. We claim that $x \in M_X$ implies $U(x) \cap D(\sim x) = \emptyset$. Indeed, $z \in U(x) \cap D(\sim x)$ implies that $\sim x \succcurlyeq z \succcurlyeq x$. But $\sim x \succcurlyeq x$ implies that x is empty, contradicting the initial selection of $x \in M_X$.

On the other hand, the condition $a \succ x$ implies $U(a) \subset U(x)$. Likewise, the contrapositive $\sim x \succ \sim a$ implies $D(\sim a) \subset D(\sim x)$. Thus, we obtain the identity $U(x) \cup D(\sim x) = X$ from $U(a) \cup D(\sim a) = X$. In summary, the subsets $U(x)$ and $D(\sim x)$ make a partition of X. However, $U(a)$ is a proper subset of $U(x)$ and $D(\sim a)$ is a proper subset of $D(\sim x)$. Therefore, $U(a)$ and $D(\sim a)$ cannot partition X, contradicting the initial assumption that element a is a Boolean hinge. □

Theorem 9.14. *Suppose X is a distributive de Morgan algebra with no self-negated objects and a is a minimal element of M_X. Then element a is a Boolean hinge if and only if $U(a)$ contains all nonempty elements of X.*

Proof. First, assume that a is a Boolean hinge. Thus, $U(a)$ and $D(\sim a)$ is a partition of X. Applying Lemma 9.1, the homomorphism $f_a : X \to \mathcal{B}$ satisfies $f_a^{-1}(\top) = U(a)$. In view of Lemma 9.2, an element x being nonempty implies $f_a(x)$ is nonempty. Thus, $f_a(x) = \top$, which implies $x \in U(a)$. We conclude that $U(a)$ contains all nonempty elements of X.

Conversely, assume $U(a)$ contains all nonempty elements of X. It follows that $D(\sim a)$ contains all empty elements of X. We first claim that the subsets $U(a)$ and $D(\sim a)$ are mutually exclusive, i. e., $U(a) \cap D(\sim a) = \emptyset$. Contrary to hypothesis, assume $x \in U(a) \cap D(\sim a)$. Then $x \succcurlyeq a$ and $\sim a \succcurlyeq x$ imply $\sim a \succcurlyeq a$, i. e., a is empty. This contradicts the initial condition $a \in M_X$.

We next claim that the subsets $U(a)$ and $D(\sim a)$ are collectively exhaustive, i. e., $U(a) \cup D(\sim a) = X$. Contrary to hypothesis, assume $x \notin U(a) \cup D(\sim a)$. Then $x \notin U(a)$ implies $a \succ x$ or $a \parallel x$, and $x \notin D(\sim a)$ likewise implies $x \succ \sim a$ or $x \parallel \sim a$. There are four possible cases to consider:

Case 1. Assume $a \succ x$ and $x \succ \sim a$. Then $a \succ \sim a$ contradicts the initial condition that $a \in M_X$.

Case 2. Assume $a \succ x$ and $x \parallel \sim a$. Since $a \succ x$ and a is a minimal element of M_X, element x is empty. Thus, $x \in D(\sim a)$, which contradicts the hypothesis that $x \parallel \sim a$.

Case 3. Assume $a \parallel x$ and $x \succ \sim a$. Since $x \succ \sim a$ and $\sim a$ is a maximal element of M_X, element x is nonempty. Thus, $x \in U(a)$, which contradicts the hypothesis that $a \parallel x$.

Case 4. Assume $a \parallel x$ and $x \parallel \sim a$. The latter condition is equivalent to $a \parallel \sim x$. Since a is a minimal element of M_X, $a \parallel x$ implies $a \wedge x$ is empty and $a \parallel \sim x$ implies $a \wedge \sim x$ is empty. Thus, $a \wedge x \in D(\sim a)$ and $a \wedge \sim x \in D(\sim a)$. Since $D(\sim a)$ is a sublattice, we obtain $(a \wedge x) \vee (a \wedge \sim x) \in D(\sim a)$. Applying the distributive law, we obtain

$$(a \wedge x) \vee (a \wedge \sim x) = a \wedge (x \vee \sim x).$$

However, the element $x \vee \sim x$ is nonempty, which implies $x \vee \sim x \in U(a)$. Thus, we obtain $a \wedge (x \vee \sim x) = a$. We thus reach the contradiction that $a \in D(\sim a)$.

Since we have reached a contradiction in all four cases, the claim that $U(a)$ and $D(\sim a)$ are collectively exhaustive is established. In summary, we have verified that the subsets $U(a)$ and $D(\sim a)$ make a partition of X. Therefore, element a is a Boolean hinge. □

Corollary. *Suppose X is a distributive de Morgan algebra with no self-negated objects. Further, suppose the infimum b of all nonempty elements of X exists. The element $a \in M_X$ is a Boolean hinge if and only if a is a minimal element of M_X such that $b \succcurlyeq a$. In particular, every atom of a Boolean algebra is a Boolean hinge.*

Proof. Since $U(a)$ is a sublattice, $U(a)$ contains all nonempty elements if and only if it contains their infimum, i. e., $b \succcurlyeq a$. In a Boolean algebra X, every element other than $\top$ or $\perp$ belongs to M_X. Therefore, the atoms of X are the same as the minimal elements of M_X. □

Lemma 9.15. *Suppose X is a finite de Morgan algebra. Let b denote the infimum of all nonempty elements of X. If b is empty, then X has no Boolean hinge.*

Proof. Assume b is empty. Contrary to hypothesis, assume $a \in X$ is a Boolean hinge. Since b is empty, we obtain $b \in D(\sim a)$. On the other hand, each nonempty element belongs to $U(a)$. Since $U(a)$ is a lattice, we obtain $b \in U(a)$. However, the Boolean hinge a must satisfy $U(a) \cap D(\sim a) = \emptyset$. This contradiction verifies that there is no Boolean hinge. □

Theorem 9.16. *Suppose X is a de Morgan algebra. If X has two Boolean hinges, i. e., $a \in X$ and $b \in X$, then the map $f : X \to \mathcal{B}^2$ defined by*

$$f(x) = (f_a(x), f_b(x))$$

is a homomorphism. This map is onto $\mathcal{B}^2$ if and only if $a \neq b$. Conversely, if X is a finite de Morgan algebra, and there exists a homomorphism from X onto $\mathcal{B}^2$, then X has at least two distinct Boolean hinges.

Proof. It is straightforward to verify that the map $f(x) = (f_a(x), f_b(x))$ preserves the algebraic operators, because f_a and f_b are homomorphisms. First,

$$f(\sim x) = (f_a(\sim x), f_b(\sim x)) = (\sim f_a(x), \sim f_b(x)) = \sim f(x).$$

Likewise,

$$(f_a(x_1 \vee x_2), f_b(x_1 \vee x_2)) = (f_a(x_1) \vee f_a(x_2), f_b(x_1) \vee f_b(x_2)),$$
$$(f_a(x_1 \wedge x_2), f_b(x_1 \wedge x_2)) = (f_a(x_1) \wedge f_a(x_2), f_b(x_1) \wedge f_b(x_2))$$

imply $f(x_1 \vee x_2) = f(x_1) \vee f(x_2)$ and $f(x_1 \wedge x_2) = f(x_1) \wedge f(x_2)$. Therefore, the map $f(x) = (f_a(x), f_b(x))$ is a homomorphism.

Next, assume that X has two distinct Boolean hinges, say $a \in X$ and $b \in X$. They must belong to M_X, because if one of them was nonempty, then there would be no middle region, wiping out the existence of the other one. Thus, the maps f_a and f_b are distinct homomorphisms onto $\mathcal{B}$. The map $f : X \to \mathcal{B}^2$ defined by $f(x) = (f_a(x), f_b(x))$ sets up the following partition:

$$\begin{aligned} f^{-1}(1,1) &= U(a) \cap U(b), \\ f^{-1}(1,0) &= U(a) \cap D(\sim b), \\ f^{-1}(0,1) &= D(\sim a) \cap U(b), \\ f^{-1}(0,0) &= D(\sim a) \cap D(\sim b). \end{aligned}$$

Notice that $\top \in f^{-1}(1,1)$ and $\bot \in f^{-1}(0,0)$. Furthermore, $x \in f^{-1}(1,0)$ if and only if $\sim b \succcurlyeq x \succcurlyeq a$. Therefore, $f^{-1}(1,0) = [a, \sim b]$ is not an empty set if and only if $\sim b \succcurlyeq a$. Applying Lemma 9.9, $b \succ a$ is impossible, because it implies b is not a minimal element of M_X. We also have $b \neq a$. Thus, $b \notin U(a)$ implies $b \in D(\sim a)$, so that $\sim b \succcurlyeq a$. Therefore, $f^{-1}(1,0)$ is not an empty set. Furthermore, we have $x \in f^{-1}(1,0)$ if and only if $\sim x \in f^{-1}(0,1)$. Therefore, $f^{-1}(0,1)$ is not an empty set. In summary, all four subsets of the partition are not empty sets. These four subsets are all sublattices of X, because the intersection of two sublattices is a sublattice.

Conversely, assume that X is finite and there exists a homomorphism $f : X \to \mathcal{B}^2$ such that f is onto. Define the projection map $\pi_1 : \mathcal{B}^2 \to \mathcal{B}$ by $\pi_1(y_1, y_2) := y_1$ and the projection map $\pi_2 : \mathcal{B}^2 \to \mathcal{B}$ by $\pi_2(y_1, y_2) = y_2$. These are homomorphisms. Let $h_i := \pi_1 \circ f$ denote the composition map for $i = 1, 2$. These are also homomorphisms. Let $a \in X$ denote the infimum of $h_1^{-1}(1)$ and let $b \in X$ denote the infimum of $h_2^{-1}(1)$. These are the Boolean hinges.

Finally, we must show that $a \neq b$. Since $a \in h_1^{-1}(1)$, we obtain $f(a) = (1,1)$ or $f(a) = (1,0)$. Yet, there exists $x \in f^{-1}(1,0)$, because $f : X \to \mathcal{B}^2$ is onto. Since $h_1(x) = 1$, we obtain $x \in U(a)$. Thus, $x \succcurlyeq a$ implies $f(x) \succcurlyeq f(a)$. Since we have $f(x) = (1,0)$, we also have $f(a) = (1,0)$. Similarly, we have $f(b) = (0,1)$. Therefore, we obtain $a \neq b$. □

Corollary. *Suppose X is a de Morgan algebra with two distinct Boolean hinges, say a and b. Then the homomorphism $f : X \to \mathcal{B}^2$ is well defined by*

$$\begin{aligned} f^{-1}(1,1) &= U(a \vee b), \\ f^{-1}(1,0) &= [a, \sim b], \\ f^{-1}(0,1) &= [b, \sim a], \\ f^{-1}(0,0) &= D(\sim a \wedge \sim b). \end{aligned}$$

Proof. The result for $f^{-1}(0,0)$ follows from Lemmas 7.1 and 7.2; the result for $f^{-1}(1,1)$ follows from Dual Lemmas 7.1 and 7.2. The other two parts are straightforward. □

Theorem 9.17. *Suppose $f : X \to Y$ is a truth-preserving homomorphism, where X is a simple truth lattice with at least five elements and Y is a de Morgan algebra. If $\{x_1, x_2, \ldots, x_n\}$ is an antichain in X, then $\{f(x_1), f(x_2), \ldots, f(x_n)\}$ is an antichain in Y such that $x_i \parallel x_j$ implies $f(x_i) \parallel f(x_j)$.*

Proof. Since f is a truth-preserving homomorphism, we obtain $f(\top) = \top$ and $f(\bot) = \bot$. Select any two distinct elements $x_i, x_j \in \{x_1, x_2, \ldots, x_n\}$. Since $x_i \vee x_j = \top$ and $x_i \wedge x_j = \bot$, we compute that

$$f(x_i) \vee f(x_j) = \top,$$
$$f(x_i) \wedge f(x_j) = \bot.$$

We claim that $f(x_i) \parallel f(x_j)$. Contrary to hypothesis, assume that $f(x_i) \succcurlyeq f(x_j)$, so that $f(x_i) = \top$ and $f(x_j) = \bot$. Select a third element $x_k \in M_X$, distinct from the other two. Since $f(x_i) \wedge f(x_k) = \bot$, we obtain $f(x_k) = \bot$. Since $f(x_k) \vee f(x_j) = \top$, we also obtain $f(x_k) = \top$. This contradiction shows that $f(x_i) \parallel f(x_j)$. Therefore, $\{f(x_1), f(x_2), \ldots, f(x_n)\}$ is an antichain in Y. □

Corollary. *Suppose X is a simple truth lattice with at least five elements and Y is a de Morgan algebra. If $f : X \to Y$ is a truth-preserving homomorphism, then it is a monomorphism (i. e., it is one-to-one).*

Proof. It is clear that $f^{-1}(\top) = \{\top\}$ and $f^{-1}(\bot) = \{\bot\}$. Assume $f(x_i) = f(x_j)$, where the elements x_i and x_j are elements of the antichain $\{x_1, x_2, \ldots, x_n\}$ in X. Since $x_i \simeq x_j$, we obtain $x_i = x_j$. □

9.5 Exercises

(1) Suppose $f : X \to Y$ is a homomorphism from de Morgan algebra X to the de Morgan algebra Y. Demonstrate that $f(x_1) \parallel f(x_2)$ implies $x_1 \parallel x_2$.

(2) Suppose X is a de Morgan algebra. Demonstrate that X^2 is a simple truth lattice if and only if X is the Boolean truth lattice $\mathcal{B}$.

(3) Suppose X is a discrete fuzzy logic with an even number of elements. Construct a homomorphism from X onto $\mathcal{B}$. Explain why there is no such homomorphism if the fuzzy logic X has an odd number of elements.

(4) Suppose X is a finite de Morgan algebra. Demonstrate that the canonical score Φ is a homomorphism $\Phi : X \to [0,1]$ if and only if the following two conditions are satisfied:
   (i) $\Phi(x \vee y) = \max\{\Phi(x), \Phi(y)\}$;
   (ii) $\Phi(x \wedge y) = \min\{\Phi(x), \Phi(y)\}$.

Demonstrate the parallel result for the standard score.

(5) Suppose X is a finite de Morgan algebra. Demonstrate that the canonical score $\Phi : X \to [0,1]$ is a homomorphism if and only if the lattice order $\succcurlyeq$ on X is total.

(6) Define the map $f : \mathcal{B}^2 \to \mathcal{P}$ by the formula $f(i,j) = (i+j)/2$. Verify that this map f is onto $\mathcal{P}$. Demonstrate that this map f is not a homomorphism without appealing to any lemmas or theorems in this chapter.

(7) Demonstrate that there is no homomorphism from $\mathcal{P}^m$ into $\mathcal{B}^n$, where m and n are natural numbers.

(8) Construct a homomorphism from $\mathcal{B}^n$ into $\mathcal{P}^n$.

(9) Suppose X is a de Morgan algebra with top $\top$ and bottom $\perp$. Demonstrate that $(\perp, 1)$ is a Boolean hinge for $X \times \mathcal{B}$.

(10) Suppose X is a de Morgan algebra with top $\top$ and bottom $\perp$. Determine two Boolean hinges for $X \times \mathcal{B}^2$. Identify the four congruence classes of the associated homomorphism onto $\mathcal{B}^2$.

10 Algebraic logic

The traditional purpose of logic is to preserve truth. The key to logical reasoning is thus conceived as explaining the meaning of logical implication and making its syntax rigorous. We will examine this problem from an algebraic perspective. This topic is vast in scope, and we will focus only on a few basic issues.

The ancient form of syllogistic reasoning conforms to the logic of sets. For example, the *categorical syllogism* takes the form

$$x \in \alpha, \quad \alpha \subseteq \beta \vdash x \in \beta.$$

The *antecedent* consists of the *minor premise* $x \in \alpha$ (e. g., Socrates is a man) and the major premise $\alpha \subseteq \beta$ (e. g., all men have two legs). The consequent is $x \in \beta$ (e. g., Socrates has two legs). The logical symbol $\vdash$ is used here to separate the antecedent and the consequent. The *disjunctive syllogism* is a more sophisticated form or reasoning involving negation:

$$x \in \alpha, \quad x \in \alpha' \cup \beta \vdash x \in \beta.$$

The major premise now takes the form $\alpha' \cup \beta$ (e. g., Socrates is not a man or has two legs). This form is the basis of material implication. The *conjunctive syllogism* can be written in the form

$$x \in \alpha, \quad x \in \alpha \cap \beta \vdash x \in \beta.$$

In this case, the major premise is $x \in \alpha \cap \beta$ (e. g., Socrates is a man and he has two legs). However, the minor premise is redundant in this case.

A more abstract version of these syllogisms can be transplanted to the de Morgan algebra. Let $p \Rightarrow q =: \sim p \vee q$ denote *material implication*. This is an asymmetric binary operator. The *disjunctive syllogism* is given by

$$p, \, p \Rightarrow q \vdash q.$$

It is interpreted as follows: p is "true" and $p \Rightarrow q$ is "true"; therefore q is "true." The simplest definition of "truth" in this context is the top $\top$. The disjunctive syllogism thus reads: $p = \top$ and $p \Rightarrow q = \top$ implies $q = \top$.

On the other hand, the *categorical syllogism* is given by

$$p, \, p \preccurlyeq q \vdash q.$$

This syllogism has a different interpretation because the major premise is a relation instead of an operator. In this case, we rely upon the defining equation for $p \preccurlyeq q$, i. e., $p \wedge q = p$. It is obvious that $p = \top$ and $p \preccurlyeq q$ implies $p \wedge q = \top$. Conversely, $p = \top$ and

https://doi.org/10.1515/9783112235812-010

$p \wedge q = \top$ imply $p \preccurlyeq q$. If we replace the premises of the categorical syllogism with p, $p \wedge q$, then it reduces to the *conjunctive syllogism*

$$p,\ p \wedge q \vdash q.$$

The simplest interpretation is that $p = \top$ and $p \wedge q = \top$ implies $q = \top$. However, we find that the conjunctive syllogism is less useful than the categorical syllogism. A key issue is the relation between material implication $p \Rightarrow q$ and the lattice order $q \succcurlyeq p$. These are two different types of implication.

More abstractly, an *implication* is a binary operator $p \rightarrowtail q$ on the algebra that has special properties that are considered essential for "sound judgment," i. e., the premises being true imply that the conclusion is true. In particular, one condition that is seldom discussed in nondual algebra involves the contrapositive. In general, the *contrapositive* of the implication $p \rightarrowtail q$ is defined as the reverse implication $\sim q \rightarrowtail \sim p$. We regard the equivalence between an implication and its contrapositive as essential to any plausible theory of logical inference. Consider material implication for example. It this case, the contrapositive $\sim q \Rightarrow \sim p$ is equivalent to $p \Rightarrow q$ provided that the law of double negation holds. Therefore, the disjunctive syllogism can be recast into the equivalent form

$$\sim q,\ \sim q \Rightarrow \sim p \vdash \sim p.$$

Now it is unclear how this should work when the algebra excludes the law of double negation as an axiom.

Many theorists advocate abandoning material implication, and the disjunctive syllogism along with it, as we plunge into nondual algebra. This approach leaves the lattice order as the most viable principle of inference. Nevertheless, one still needs the law of polarity to make complete sense out of the categorical syllogism. In essence, this law indicates that the precedent p is the "reason" for q. Thus, the absence of q should imply that the reason p for q is absent. In Buddhist logic, the reason usually refers to the causes and conditions. Several prominent nondual algebras weaken this law to such an extent that $\sim q \succcurlyeq \sim p$ does not always imply $p \succcurlyeq q$. How can one claim that p is the "reason" for q in such a situation? An analysis of the law of polarity is vital to addressing these concerns.

10.1 Material implication

Classical logic is committed to material implication $p \Rightarrow q$ as its main instrument of logical inference. In the framework of the Boolean algebra, the following result is well known:

$$\sim p \vee q = \top \quad \text{if and only if } q \succcurlyeq p.$$

This elementary result theoretically justifies the premise that $p \Rightarrow q$ is "true." In essence, we can rely upon our knowledge of the lattice order to verify its truth. Yet, it does not generalize *prima facie* to the de Morgan algebra. In particular, suppose $q \succcurlyeq p$ implies $\sim p \vee q = \top$ in a particular de Morgan algebra. Then the selection $q = p$ implies $\sim p \vee p = \top$. Therefore, the algebra satisfies the law of excluded middle, i. e., it is a Boolean algebra. This is obviously a very restrictive condition. We must thus limit its scope of application when we work in the de Morgan algebra.

Lemma 10.1. *Suppose X is a distributive de Morgan algebra. If* $p \vee \sim p = \top$, *then* $q \succcurlyeq p$ *if and only if* $\sim p \vee q = \top$.

Of course, this result does not fully explain why the categorical and disjunctive syllogisms are both sound inferences in the de Morgan algebra. We can only apply it if the antecedent p belongs to the Boolean center of the de Morgan algebra.

Example. The Belnap logical lattice is represented by the Hasse diagram in Figure 10.1. We rely here upon the power set representation. The involution is determined by interchanging the marks "T" and "F". Thus, the top and bottom are opposites, but the two medians are self-negated. We now take $p = \{T, F\}$ and $q = \{\} = \emptyset$. Notice that $\sim p \vee q = \{T\}$ and $p \parallel q$. In other words, $p \Rightarrow q$ being "true" does not necessarily imply $q \succcurlyeq p$.

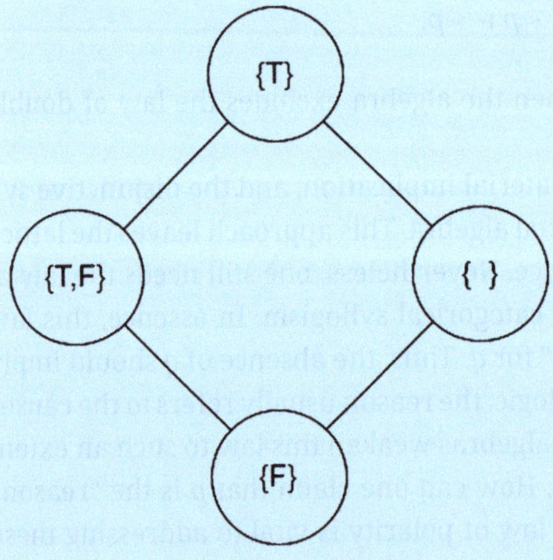

Figure 10.1: The Belnap logical lattice.

The reasoning behind logical inference can also be presented as a syllogism. The abstract form of the *categorical syllogism* is given by $p, q \succcurlyeq p \vdash q$. The standard truth-functional interpretation of this formula is that p is true and $q \succcurlyeq p$ is true, therefore q is true. This chain of reasoning is easy to verify in the de Morgan algebra. By definition, $q \succcurlyeq p$ if and only if $q = p \vee q$. The substitution of $p = \top$ into the equation $q = p \vee q$ yields $q = \top$. Likewise, the abstract form of the disjunctive syllogism is given by $p, p \Rightarrow q \vdash q$. The substitution of $p = \top$ into the form $\sim p \vee q = \top$ reduces it to $q = \top$. Both forms of the syllogism seem to preserve truth, but there is a subtle shift in the meaning of "truth" between the two conjunctive and disjunctive forms. The latter method fully relies upon assigning the truth-value of $\top$ to both premises. The former method relies upon the "truth" of the equation $q = p \vee q$. This type of truth is not the same as truth-value in a de Morgan truth lattice. In conclusion, both types of inference work in the de Morgan algebra, but their connection is not apparent.

One way to resolve this dilemma is to expand the notion of "truth." A common method in modern logic is to designate a special subset of truth-values as "truth." Whatever is selected as "truth," it should be an up-set of the truth lattice. Moreover, the involution of "truth" should be "false." We tentatively suggest that one should include nonempty elements as truth and empty elements as falsehood. Notice that this interpretation reduces to the conventional one in a Boolean algebra X, because $\top$ is the only nonempty element and $\bot$ is the only empty element in X.

Example. Let us return to the Belnap logical lattice analyzed in the previous example. Suppose the nonempty elements $\{\{T\}, \{T, F\}, \emptyset\}$ are designated as "truth" and the empty elements $\{\{F\}, \{T, F\}, \emptyset\}$ are designated as "false." Thus, the two medians are both "true" and "false." Notice that $\sim p \vee q = \{F\}$ if and only if $p = \{T\}$ and $q = \{F\}$. Therefore, $\sim p \vee q$ being nonempty implies q is nonempty. In summary, the disjunctive syllogism is sound with the Belnap logical lattice and the designation of "nonemptiness is truth."

One of perceived semantic flaws of material implication is the notion that a false premise allows one to deduce any proposition, whether it is true or false, as a conclusion. This type of logic is said to be *explosive*. In contrast, the logical system is said to be *paraconsistent* provided that it is not explosive [18]. The anomaly of explosiveness is associated with logical implications of the generic form

$$p, \sim p \vdash q.$$

The special symbol "$\vdash$" denotes a binary relation connecting the premise $(p, \sim p)$ to the conclusion q. We interpret the premise here as $p \wedge \sim p$ in the propositional calculus. The first step in the general argument for explosiveness is to assume that the premise $p \wedge \sim p$ is "true." An argument that begins with a contradiction is obviously non-Boolean. The second step in the general argument is to deduce that p is "true" and $\sim p$ is "true." An implicit assumption here is that the category of "true" propositions is an up-set. Moreover, one deduces that $\sim p \vee q$ is "true," because $\sim p$ is "true." The third step in the argument is to apply the disjunctive syllogism: Since p is "true" and $\sim p \vee q$ is "true," the conclusion is that q is "true." Curiously, the same argument also shows that $\sim q$ is "true." In other words, everything is "true" all at once. This chain of reasoning is clearly designed to challenge conventional logic.

Let us examine this scheme in the framework of the de Morgan algebra. The first step is the premise that $p \wedge \sim p$ is "true." Nevertheless, $p \wedge \sim p$ has the algebraic form of an empty proposition, which is associated with falsehood. Therefore, the proposition

$p \wedge \sim p$ must be both true and false. In other words, it is both empty and nonempty. The only type of proposition that fits this description is a self-negated one (see Lemma 4.4). The next result is elementary.

Lemma 10.2. *Suppose X is a de Morgan algebra. The object $a \wedge \sim a$ is self-negated if and only if the object a itself is self-negated.*

Of course, a logician might object to the interpretation of the comma in "$p, \sim p$" as a conjunction, but that is not a serious objection. The alternative is to interpret the premise p as both true and false. Thus, it is self-negated. The next two steps assume that "truth" is an up-set. This way of thinking is consistent with our working paradigm: "nonemptiness is truth and emptiness is falsehood." The last step is the evocation of material implication. Nevertheless, there are many cases where this type of reasoning is invalid.

Example. Let us revisit the pentalemma with a new labeling scheme on display in Figure 10.2.

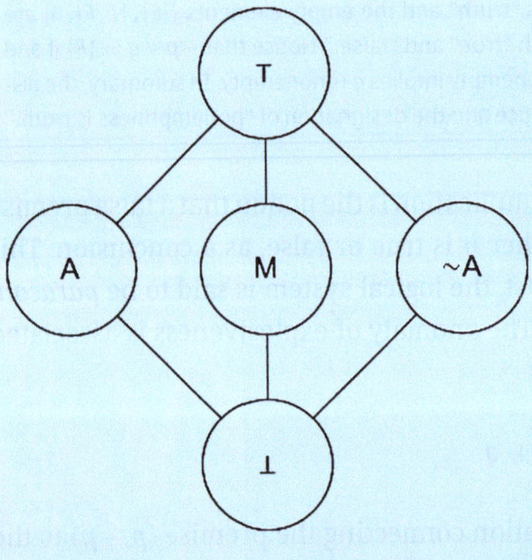

Figure 10.2: The Hasse diagram of the pentalemma.

It is understood that the object M in the center is self-negating. Notice that both M and $\top$ are nonempty and thus have a claim to truth. However, only the top $\top$ is nonempty-only. Let $p = M$ and $q = A$. Then $\sim p \vee q = \top$ and $p \parallel q$. If we were to accept the disjunctive syllogism with material implication, then we would falsely conclude $q = \top$.

The only de Morgan algebra that structurally eliminates the analytic difficulties associated with $p \parallel q$ is fuzzy logic. Suppose X is a fuzzy logic. Since the canonical score defines an isomorphism in this case, we may assume that $X = [0, 1]$ without loss of generality. Thus, the element $x \in X$ is self-negated if and only if $x = 1/2$; x is nonempty-only if and only if $x > 1/2$; and x is empty-only if and only if $x < 1/2$. We identify nonemptiness with truth and emptiness with falsehood. The middle value 1/2 is thus both true and false. In order to pinpoint the behavior of material implication, we will compute

$\sim p \vee q = \max\{1-p, q\}$. Let us further interpret the region $\{x \in [0,1] : x > 1/2\}$ as "truth" and $\{x \in [0,1] : x < 1/2\}$ as "falsehood." In other words, truth consists of the nonempty-only objects and falsehood consists of the empty-only objects. The middle value 1/2 is both "true" and "false." The next simple result drives the disjunctive syllogism.

Lemma 10.3. *Suppose X is a fuzzy logic. Then* $\sim p \vee q > 1/2$ *if and only if* $q > 1/2$ *or* $p < 1/2$.

The utter mathematical triviality of this lemma is actually a source of strength in making inferences with it. The following two versions of the disjunctive syllogism are sound in a fuzzy logic:

(1) $(p > 1/2), (\sim p \vee q > 1/2) \vdash (q > 1/2)$;
(2) $(p \geq 1/2), (\sim p \vee q > 1/2) \vdash (q > 1/2)$.

Ordinarily, one takes $p > 1/2$ as the minor premise. This property indicates that p is "true." Likewise, the major premise $\sim p \vee q$ is "true." Therefore, the conclusion q is" true." However, the syllogism remains sound when $p = 1/2$. In other words, the second formulation of the disjunctive syllogism is sound when the minor premise is only "half-true." This property verifies paraconsistency. When we apply our arguments directly to the logic of paradox, we arrive at conclusions similar to those of Graham Priest [19].

The basic paradigm behind this construction is that "nonemptiness is truth and emptiness is falsehood." However, we have not yet considered the possibility that the major premise is both "true" and "false."

Corollary. *Suppose X is a fuzzy logic. Then* $\sim p \vee q = 1/2$ *if and only if the following three conditions are met:*

(1) $p \geq 1/2$;
(2) $q \leq 1/2$;
(3) $p = 1/2$ *or* $q = 1/2$.

This result suggests that a "half-true" major premise can lead to unsound judgment. For example, if the minor premise is half-true (i. e., $p = 1/2$) and the major premise is half-true (i. e., $\sim p \vee q = 1/2$), then the conclusion can be false. For instance, if we select $p = 1/2$ and $q = 0$, then we obtain $\sim p \vee q = 1/2$. On the other hand, if the minor premise is true (i. e., $p > 1/2$) and the major premise is half-true (i. e., $\sim p \vee q = 1/2$), then the conclusion is half-true (i. e., $q = 1/2$). This type of anomaly is pervasive.

Example. Suppose X is a de Morgan algebra with a self-negated element m. Select $p = m$ and $q = \bot$. We compute that $\sim p \vee q = m$. Therefore, p is nonempty; $\sim p \vee q$ is nonempty; and q is empty-only.

Example. Suppose X is a de Morgan algebra with a self-negated element m. Further suppose that exist a pair of empty-only elements, say a and b, such that $a \parallel b$ and $a \vee b = m$. (These algebras are not hard to find.) Select $p = \sim a$ and $q = b$. Then p is nonempty-only; $\sim p \vee q$ is nonempty; and q is empty-only.

In comparison, the conjunctive syllogism is more robust than the disjunctive syllogism. In particular, p is nonempty and $q \succcurlyeq p$ imply q is nonempty. Moreover, p is nonempty-only and $q \succcurlyeq p$ imply q is nonempty-only. Thus, we can designate “nonemptiness as truth” or “nonemptiness-only as truth” without any complications.

10.2 The polarity

An important method for evaluating the axioms of the de Morgan algebra is to study them in the context of a larger algebraic framework. Let us reinvestigate the status of the law of polarity without reliance upon the law of double negation. We need to distinguish two other versions of this law here:

(*Weak law of polarity*) $x \succcurlyeq y$ implies $f(y) \succcurlyeq f(x)$;

(*Strong law of polarity*) $x \succcurlyeq y$ if and only if $f(y) \succcurlyeq f(x)$.

A surjective map f that satisfies the strong law is a dual automorphism. The strong law obviously implies the weak law. Furthermore, the strong law can be partitioned into symmetric and asymmetric components of the lattice order:

(1) $x = y$ if and only if $f(y) = f(x)$;

(2) $x \succ y$ if and only if $f(y) \succ f(x)$.

The former condition asserts that f is one-to-one, and the latter condition is what we previously called the “law of polarity.” If the law of double negation prevails, then this law is obviously equivalent to the strong law. That is why we never drew a clear distinction until now. We next investigate the consequences of the weak law of polarity.

Lemma 10.4. *The following three conditions on a lattice X with unary operator $f : X \to X$ are pairwise equivalent:*

(1) *$x \succcurlyeq y$ implies $f(y) \succcurlyeq f(x)$;*

(2) *$f(x) \wedge f(y) \succcurlyeq f(x \vee y)$;*

(3) *$f(x \wedge y) \succcurlyeq f(x) \vee f(y)$.*

Condition (1) is the weak law of polarity. On the other hand, conditions (2) and (3) are weak versions of the de Morgan laws. This result justifies the nomenclature that the unary operator $f : X \to X$ described by condition (1) is called a *polarity*.

We adopt the following standard notation: f^0 is the identity map; $f^1 = f$ and $f^{n+1} = f \circ f^n$ for every natural number n. The next simple result is implicit in Blyth and Varlet [20]. The important point is that it depends only on the weak law of polarity.

Lemma 10.5. *Suppose the lattice X is equipped with the polarity $f : X \to X$. If $f^2 \succcurlyeq f^0$, then $f^3 = f$.*

The *skeleton* of the lattice X with polarity $f : X \to X$ is defined by

$$S(X) := \{f(x) : x \in X\}.$$

The restricted map $f : S(X) \to S(X)$ is obviously well defined.

Corollary. *Suppose the lattice X is equipped with the polarity* $f : X \to X$. *If* $f^2 \succcurlyeq f^0$, *then* $f^2(x) = x$ *for every* $x \in S(X)$.

We thus recover the law of double negation whenever the polarity is restricted to the skeleton. Furthermore, the polarity $f : S(X) \to S(X)$ is invertible. Therefore, it satisfies the strong law of polarity. We conclude that $S(X)$ is a de Morgan algebra.

Example. Consider the lattice with the Hasse diagram in Figure 10.3.

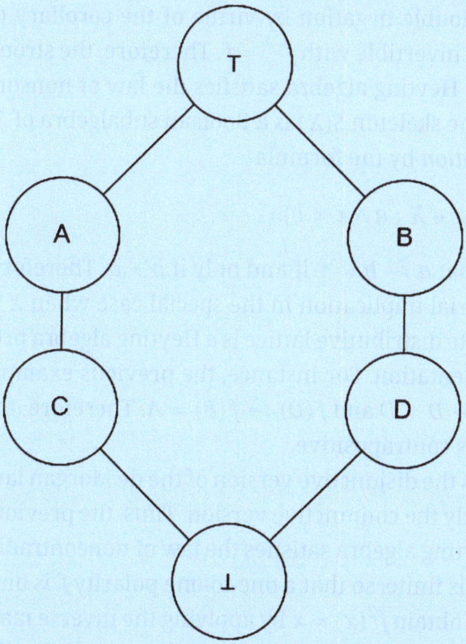

Figure 10.3: The Hasse diagram of a simple Heyting algebra.

The polarity $f : X \to X$ is well defined by the second row in the next table. Brute force computations (e. g., truth tables) reveal that the de Morgan laws are satisfied.

x	$\top$	A	B	C	D	$\bot$
$f(x)$	$\bot$	B	A	B	A	$\top$
$f^2(x)$	$\top$	A	B	A	B	$\bot$
$f^3(x)$	$\bot$	B	A	B	A	$\top$

Moreover, the above table confirms that $f^2(x) \succcurlyeq x$ and $f^3(x) = f(x)$. The skeleton $S(X)$ consists of the elements $\{\top, A, B, \bot\}$, which is a tetralemma.

Next consider a classic alternative to complementation. Suppose X is a lattice. We say that the element $a \in X$ is *pseudocomplemented* provided that the element

$$a^\star := \max\{x \in X : a \wedge x = \bot\}$$

exists, in which case we can define $a^\star$ as the *pseudocomplement* of element a. Notice that $a^\star \wedge a = \bot$, i. e., pseudocomplementation does not allow contradictions. It is well known that the pseudocomplement satisfies the identity $a^{\star\star} \succcurlyeq a$.

Case study. The *Heyting algebra* defines negation as pseudocomplementation. It also satisfies the de Morgan law for disjunctions, i. e., $f(x \vee y) = f(x) \wedge f(y)$. In view of Lemma 10.3, it satisfies the weak law of polarity. Furthermore, the restricted polarity $f : S(X) \to S(X)$ satisfies the law of double negation by virtue of the corollary to Lemma 10.5. The restricted polarity is thus invertible with $f^{-1} = f$. Therefore, the strong law of polarity prevails on $S(X)$. Since the Heyting algebra satisfies the law of noncontradiction, i. e., $x \wedge f(x) = \bot$; we find that the skeleton $S(X)$ is a Boolean subalgebra of X.

The Heyting algebra X defines *implication* by the formula

$$a \rightarrowtail b := \max\{x \in X : a \wedge x \preccurlyeq b\}.$$

This operator has the intuitive property that $a \rightarrowtail b = \top$ if and only if $b \succcurlyeq a$. Therefore, Heyting implication is equivalent to material implication in the special case when X is Boolean. According to Blyth [21], every finite distributive lattice is a Heyting algebra provided that the polarity is pseudocomplementation. For instance, the previous example is a finite Heyting algebra. Notice that $B \rightarrowtail D = D$ and $f(D) \rightarrowtail f(B) = A$. Therefore, the Heyting implication is not equivalent to its contrapositive.

The Heyting algebra X always satisfies the disjunctive version of the de Morgan law, i. e., $f(x \vee y) = f(x) \wedge f(y)$; but not necessarily the conjunctive version. Thus, the previous lemma applies. We also know that the Heyting algebra satisfies the law of noncontradiction, i. e., $x \wedge f(x) = \bot$. Further, suppose X is finite, so that a one-to-one polarity f is onto and thus invertible. Since $f^3(x) = f(x)$, we obtain $f^2(x) = x$ by applying the inverse map, i. e., the law of double negation holds. Therefore, a finite Heyting algebra satisfies the strong law of polarity if and only if it is a Boolean algebra.

10.3 Ockham algebras

An Ockham algebra provides a platform for removing the law of double negation without modifying the de Morgan laws. By definition, an *Ockham algebra* is a bounded (distributive) lattice equipped with a unary operator $f : X \to X$ satisfying the following properties:

(1) $f(\top) = \bot$ and $f(\bot) = \top$;
(2) $f(x_1 \vee x_2) = f(x_1) \wedge f(x_2)$;
(3) $f(x_1 \wedge x_2) = f(x_1) \vee f(x_2)$.

The first condition asserts that the top and bottom are opposites; the second and third conditions are the de Morgan laws. What is missing is the law of double negation. Thus, an Ockham algebra is a generalization of the de Morgan algebra. Although Ockham algebras are sometimes assumed to be distributive, this extra property plays only a minor role in the analysis developed here.

Case study. The *MS-algebra* is an Ockham algebra X such that $f^2(x) \succcurlyeq x$ for every $x \in X$. As a special case, the Stone algebra is the MS-algebra that defines the polarity as pseudo-complementation, i. e., $f(x) = x^*$. The previous example is a Stone algebra. In general, the MS-algebra satisfies the identity $f^3 = f$ and its skeleton is a de Morgan algebra. These algebras have been thoroughly studied by Blyth and Varlet [22].

An Ockham algebra always satisfies the weak law of polarity. We next investigate the extent to which it obeys the strong law.

Lemma 10.6. *Suppose X is a lattice with a polarity $f : X \to X$ that satisfies at least one de Morgan law. Then the polarity f satisfies the strong law of polarity if and only if it is one-to-one.*

For example, the Heyting algebra satisfies the disjunctive de Morgan law, i. e., $f(x \vee y) = f(x) \wedge f(y)$; but not necessarily the conjunctive version. In this case, the polarity is not one-to-one as we confirm below.

Example. Figure 10.4 is the Cartesian graph of a real-valued function. It illustrates why a polarity does not necessarily satisfy the strong law of polarity.

The function $f : [0,1] \to [0,1]$ satisfies the weak law of polarity, i. e., it is monotone decreasing. If we select $x > y$ in the horizontal region of the graph, however, we obtain $f(x) = f(y)$. Thus, it violates the strong law of polarity.

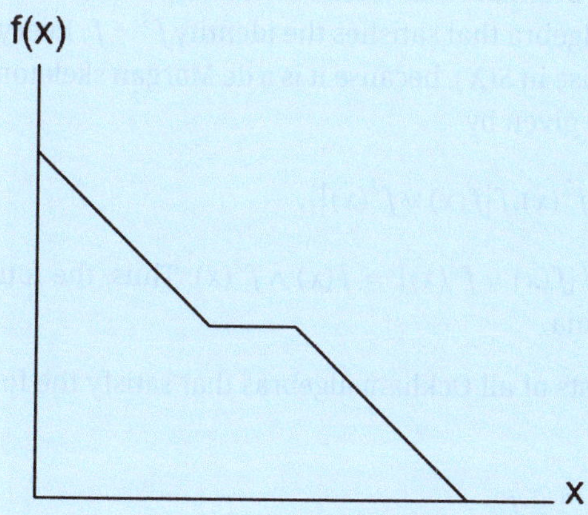

Figure 10.4: The graph of a monotone decreasing real-valued function.

Theorem 10.7. *Suppose $f : X \to X$ is a unary operator onto the lattice X. If the strong law of polarity holds, then X is an Ockham algebra.*

For example, a Heyting algebra that is not an Ockham algebra must violate the strong law of polarity. This result utilizes the surjective property of f only to establish that the top and bottom are opposites. If the lattice X is finite, then the map f is onto if and only if it is one-to-one. We can therefore omit the assumption that f is onto in the finite case. Likewise, if we restrict the polarity f to the skeleton $S(X)$, then the assumption of surjectivity is redundant.

Case study. Let us briefly study how Buddhist logic works in an Ockham algebra. The four premises of the tetralemma are given by

$$\{x, f(x), x \vee f(x), f[x \vee f(x)]\}.$$

The last one in this list, i. e., the negated disjunction, is equivalent to $f(x) \wedge f^2(x)$. This identity can be interpreted as a new version of Nagarjuna's identity. We next define "emptiness" broadly as the negated disjunction of all four premises. The following basic result is relevant.

Lemma 10.8. *If X is a distributive Ockham algebra with polarity $f : X \to X$, then $x \vee f(x) \succcurlyeq f(x) \wedge f^2(x)$.*

It immediately follows from this lemma that the disjunction of the four premises is given by $x \vee f(x)$, i. e., "both existence and nonexistence," and the negated disjunction of the four premises is given by $f[x \vee f(x)]$, i. e., "neither existence nor nonexistence." The definition of "emptiness" has therefore not substantially changed. However, "neither" is not necessarily equivalent to the "oneness of existence and nonexistence," i. e., $x \wedge f(x)$. In other words, Nagarjuna's identity in its original form can be violated.

Let us next specialize to an Ockham algebra that satisfies the identity $f^3 = f$. Everything is simplified if we begin with a premise in $S(X)$, because it is a de Morgan skeleton. In this case, the four premises for $f(x)$ are given by

$$\{f(x), f^2(x), f(x) \vee f^2(x), f[f(x) \vee f^2(x)]\},$$

and Nagarjuna's identity takes the form $f[f(x) \vee f^2(x)] = f(x) \wedge f^2(x)$. Thus, the four premises can be arranged into a tetralemma.

By definition, the *Berman class* consists of all Ockham algebras that satisfy the following identity:

$$f^{2m+n} = f^n$$

for fixed integers $m \geq 1$ and $n \geq 0$. Let $B(m, n)$ denote the Berman class for parameters m and n. For example, every MS-algebra belongs to the Berman class $B(1, 1)$, because it satisfies the identity $f^3 = f$. It is also well known that every finite Ockham algebra belongs to a Berman class [23]. We next extend the concept of the skeleton to accommodate the general Berman class. Indeed, if X belongs to the Berman class $B(m,n)$, then the *Berman skeleton* of X is defined by $S_n(X) := \{f^n(x) : x \in X\}$. We previously discussed the special case $B(1,1)$. Notice that $S_n(X) = S(X)$ provided that $n = 1$. This class of algebras satisfies the identity $f^{2m+1} = f$ for some $m \geq 1$.

Lemma 10.9. *Suppose X is an Ockham algebra in the Berman class $B(m,n)$. Then its Berman skeleton $S_n(X)$ is an Ockham subalgebra of X in the Berman class $B(m,0)$.*

Let us analyze the structure of the Berman class $B(m,0)$ in more depth. We find that these algebras behave more elegantly than the other Ockham algebras.

Theorem 10.10. *Suppose X is an Ockham algebra in the Berman class $B(m,0)$. Then the polarity $f : X \to X$ is invertible and the strong law of polarity holds.*

The defining property of the Berman class $B(m,0)$ is the existence of a natural number $m \geq 1$ such that the identity $f^{2m} = f^0$ holds for all $x \in X$. However, each element in X can have its own value $k \geq 1$ such that $f^k(x) = x$ holds. By definition, the sequence $\{a_i\}_{i=1}^k$ is a *cycle* provided that $a_i \neq a_j$ for every $i \neq j$ and

$$f(a_i) = \begin{cases} a_{i+1} & \text{if } 1 \leq i < k, \\ a_1 & \text{if } i = k. \end{cases}$$

The number k of the cycle $\{a_i\}_{i=1}^k$ is called its *length*. The length k is thus the smallest natural number such that $f^k(a_1) = a_1$. This property is also true for all elements in the cycle $\{a_i\}_{i=1}^k$. In particular, the set $\{\top, \bot\}$ can be arranged into a cycle of length 2. This cycle is always present in any Ockham algebra.

Lemma 10.11. *Suppose X belongs to the Berman class $B(m,0)$. Every element $x \in X$ belongs to a cycle, and the least common multiple $2k$ of the lengths of the cycles corresponds to the least parameter such that X belongs to $B(k,0)$. Furthermore, the parameter m is a multiple of k.*

Example. Consider the lattice with the Hasse diagram displayed in Figure 10.5.

Although it looks like the pentalemma, its polarity is defined much differently. The polarity is defined in the following table:

x	$\top$	a	b	c	$\bot$
$f(x)$	$\bot$	b	c	a	$\top$

The de Morgan laws can be verified straightforwardly. This polarity thus defines an Ockham algebra. Notice that the subset $\{\top, \bot\}$ is a cycle of length 2 and the subset $\{a, b, c\}$ is a cycle of length 3. Therefore, this algebra belongs to $B(3,0)$.

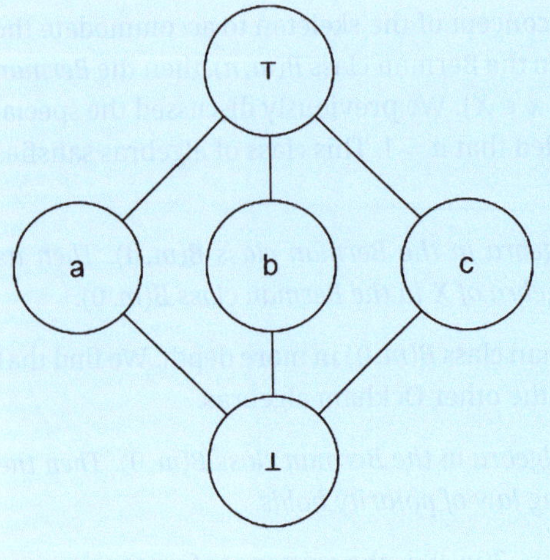

Figure 10.5: An Ockham algebra of the Berman class $B(3, 0)$.

Lemma 10.12. *Suppose X is an Ockham algebra of the Berman class $B(m, n)$. If the polarity $f : X \to X$ is one-to-one, then it is onto and thus invertible.*

If the polarity is invertible, then it immediately follows that

$$X = S(X) = S_n(X).$$

According to Theorem 10.7, X belongs to the Berman class $B(m, 0)$. This result also highlights the role of the strong law of polarity. It inevitably fails in a Berman class such that $X \neq S(X)$. If the strong law fails, then the idea that the lattice order can be regarded as an "implication" is weakened.

Theorem 10.13. *Suppose X belongs to the Berman class with parameters m and n, where m is an odd number. Then the map $f^m : S_n(X) \to S_n(X)$ is a well-defined polarity such that $S_n(X)$ is a de Morgan algebra with respect to f^m.*

For example, consider the special case when $m \geq 1$ and $n = 0$, i. e., the polarity satisfies the identity $f^{2m} = f^0$. We immediately obtain $S_0(X) = X$. If m is odd, then the above theorem applies.

One can also say something more about the structure of the element $f^m(x)$, where m is odd. If the element $x \in X$ belongs to a cycle with odd length, i. e., it divides m, then we obtain $f^m(x) = x$. Thus, element x is self-negated with respect to the involution f^m. On the other hand, if the element $x \in X$ belongs to a cycle with even length, i. e., it does not divide m, then $f^m(x) \neq x$. Thus, the element x is not self-negated with respect to the involution f^m. The parity of the length of the cycle thus provides information about whether or not $\{x, f^m(x)\}$ are opposites.

Example. The de Morgan algebra belongs to the Berman class $B(1,0)$. Thus, every element belongs to a cycle of length 1 or 2. The next relevant Berman class is $B(3,0)$. In this case, every element belongs to a cycle of length that is a divisor of 6. Moreover, there is at least one cycle of length 3 or 6.

The previous theorem cannot be applied when there exists a cycle with length equal to a multiple of 4. In this case, the parameter m is an even number, which renders the key map $f^m : S_n(X) \to S_n(X)$ into a lattice automorphism, i. e., it preserves the lattice operators:

$$f^m(x \vee y) = f^m(x) \vee f^m(y);$$
$$f^m(x \wedge y) = f^m(x) \wedge f^m(y).$$

Since $f^{2m} = f^0$ on $S_n(X)$, the lattice automorphism f^m is idempotent. The subalgebra $S_n(X)$ thus exhibits left-right symmetry. If we further assume that X is finite, then Lemma 5.8 applies. Therefore, we obtain $f^m(x) \simeq x$ for every $x \in S_n(X)$.

Assuming the parameter m is odd, the map $f^m : S_n(X) \to S_n(X)$ is a dual automorphism. Thus, we can interpret f^m as a new polarity. When we switch to this map as the designated involution, then the polarity f plays a secondary role. We broadly interpret an iteration of f as a phase change from an object x toward its opposite $f^m(x)$. Thus, m phase changes reaches always the opposite. Theorem 10.13 can be generalized further. For every real number x, let $\lceil x \rceil$ denote the smallest integer k such that $k \geq x$.

Corollary. *If X belongs to the Berman class $B(m,n)$ with respect to the polarity f, where m is odd, then X belongs to the Berman class $B(1,k)$ with respect to the polarity f^m, where $k = \lceil n/m \rceil$.*

For example, further suppose that $0 < n \leq m$. Then we find that X belongs to the Berman class $B(1,1)$ with respect to the polarity f^m.

10.4 Proofs to this chapter

Lemma 10.1. *Suppose X is a distributive de Morgan algebra. If $p \vee {\sim} p = \top$, then $q \succcurlyeq p$ if and only if ${\sim} p \vee q = \top$.*

Proof. Assume $p \vee {\sim} p = \top$ throughout this argument. Further, assume $q \succcurlyeq p$, i. e., $q = p \vee q$. Then we directly compute that

$${\sim} p \vee q = {\sim} p \vee (p \vee q) = ({\sim} p \vee p) \vee q = \top.$$

Conversely, assume ${\sim} p \vee q = \top$. Then we obtain $p \wedge ({\sim} p \vee q) = p$. Applying the distributive law, we directly compute that

$$p \wedge ({\sim} p \vee q) = (p \wedge {\sim} p) \vee (p \wedge q) = p \wedge q.$$

Since $p = p \wedge q$, we obtain $q \succcurlyeq p$. □

Lemma 10.2. *Suppose X is a de Morgan algebra. The object $a \wedge \sim a$ is self-negated if and only if the object a itself is self-negated.*

Proof. Assume $a \wedge \sim a = \sim(a \wedge \sim a) = a \vee \sim a$. Since $a \vee \sim a \succcurlyeq a \succcurlyeq a \wedge \sim a$, we obtain $a = a \wedge \sim a$. Thus, a is self-negated. Conversely, assume $a = \sim a$. Then we obtain $a \wedge \sim a = a \wedge a = a$. Thus, $a \wedge \sim a$ is self-negated. □

Lemma 10.3. *Suppose X is a fuzzy logic. Then $\sim p \vee q > 1/2$ if and only if $q > 1/2$ or $p < 1/2$.*

Proof. The involution for this fuzzy logic is defined by $\sim p := 1 - p$. Thus, we obtain $\sim p \vee q = \max\{1 - p, q\}$. It immediately follows that $\sim p \vee q > 1/2$ if and only if $p < 1/2$ or $q > 1/2$. □

Corollary. *Suppose X is a fuzzy logic. Then $\sim p \vee q = 1/2$ if and only if the following three conditions are met:*
(1) $p \geq 1/2$;
(2) $q \leq 1/2$;
(3) $p = 1/2$ *or* $q = 1/2$.

Proof. Assume $\sim p \vee q = 1/2$. Then we obtain $\sim p \leq 1/2$ and $q \leq 1/2$. The former condition is equivalent to $p \succcurlyeq 1/2$. However, $\sim p < 1/2$ and $q < 1/2$ imply $\sim p \vee q < 1/2$, a contradiction. Therefore, $p = 1/2$ or $q = 1/2$.

Conversely, assume the three conditions hold. Then $\sim p \leq 1/2$ and $q \leq 1/2$ imply $\sim p \vee q \leq 1/2$. Furthermore, $p = 1/2$ or $q = 1/2$ implies $\sim p \vee q \geq 1/2$. Therefore, we obtain $\sim p \vee q = 1/2$. □

Lemma 10.4. *The following three conditions on a lattice X with unary operator $f : X \to X$ are pairwise equivalent:*
(1) $x \succcurlyeq y$ *implies* $f(y) \succcurlyeq f(x)$;
(2) $f(x) \wedge f(y) \succcurlyeq f(x \vee y)$;
(3) $f(x \wedge y) \succcurlyeq f(x) \vee f(y)$.

Proof. Assume condition (1) holds. Then $x \vee y \succcurlyeq x$ implies $f(x) \succcurlyeq f(x \vee y)$ and $x \vee y \succcurlyeq y$ implies $f(y) \succcurlyeq f(x \vee y)$. Thus, condition (2) holds. Conversely, assume condition (2) holds. Then the condition $x \succcurlyeq y$ yields $x = x \vee y$, so that $f(x) \wedge f(y) \succcurlyeq f(x)$ by virtue of condition (2). Since $f(x) \succcurlyeq f(x) \wedge f(y)$ always holds, we obtain $f(x) = f(x) \wedge f(y)$ by antisymmetry. Therefore, $f(y) \succcurlyeq f(x)$, which verifies condition (1). The equivalence between conditions (1) and (3) follows by a dual argument. □

Lemma 10.5. *Suppose the lattice X is equipped with the polarity $f : X \to X$. If $f^2 \succcurlyeq f^0$, then $f^3 = f$.*

Proof. Assume $f^2(x) \succcurlyeq x$ for every $x \in X$. Let $y = f(x)$. Then the relation $f^2(y) \succcurlyeq y$ follows by direct substitution. Thus, we obtain $f^3(x) \succcurlyeq f(x)$. Since f is a polarity, we further obtain $f^3(x) \preccurlyeq f(x)$. Therefore, we conclude that $f^3 = f$. □

Corollary. *Suppose the lattice X is equipped with the polarity* $f : X \to X$. *If* $f^2 \succcurlyeq f^0$, *then* $f^2(x) = x$ *for every* $x \in S(X)$.

Proof. Let $y = f(x)$. Then we obtain $f^2(y) = f^3(x) = f(x) = y$. □

Lemma 10.6. *Suppose X is a lattice with a polarity* $f : X \to X$ *that satisfies at least one de Morgan law. Then the polarity f satisfies the strong law of polarity if and only if it is one-to-one.*

Proof. First, assume that the strong law of polarity holds. Then $f(x) = f(y)$ implies $f(x) \succcurlyeq f(y)$ and $f(y) \succcurlyeq f(x)$, so that $y \succcurlyeq x$ and $x \succcurlyeq y$. So, we obtain $x = y$ by virtue of antisymmetry. Therefore, $f : X \to X$ is one-to-one. Conversely, assume that $f : X \to X$ is one-to-one. Suppose that $f(y) \succcurlyeq f(x)$. We must consider two cases, depending on which de Morgan law is operative.

Case 1. Assume the de Morgan law for conjunctions is operative. Then we obtain $f(y) = f(x) \vee f(y) = f(x \wedge y)$. Since $f : X \to X$ is one-to-one, it follows that $y = x \wedge y$. We thus obtain $x \succcurlyeq y$. Therefore, the strong law holds in this case.

Case 2. Assume the de Morgan law for disjunctions is operative. Then we obtain $f(x) = f(x) \wedge f(y) = f(x \vee y)$. Since $f : X \to X$ is one-to-one, it follows that $x = x \vee y$. We thus obtain $x \succcurlyeq y$. Therefore, the strong law holds in this case as well. □

Theorem 10.7. *Suppose* $f : X \to X$ *is a unary operator onto the lattice X. If the strong law of polarity holds, then X is an Ockham algebra.*

Proof. Assume that the strong law of polarity holds. Notice that $f(x) = f(y)$ implies $f(x) \succcurlyeq f(y)$ and $f(y) \succcurlyeq f(x)$. Thus, we obtain $y \succcurlyeq x$ and $x \succcurlyeq y$. Since the lattice order $\succcurlyeq$ is antisymmetric, we find that $x = y$. Therefore, f is one-to-one. Since f is one-to-one, we also obtain $f(\alpha \cap \beta) = f(\alpha) \cap f(\beta)$ for arbitrary subsets α and β of X. Furthermore, the identity $f(\inf \alpha) = \sup f(\alpha)$ follows directly from the strong law of polarity.

We next derive the de Morgan law for disjunctions. We directly calculate from the strong law that

$$f(x \vee y) = f(\inf[U(x) \cap U(y)]) = \sup f[U(x) \cap U(y)]$$

and

$$f(x) \wedge f(y) = \sup\{D[f(x)] \cap D[f(y)]\}.$$

Since f is one-to-one, we further obtain

$$f[U(x) \cap U(y)] = f[U(x)] \cap f[U(y)] = D[f(x)] \cap D[f(y)].$$

Therefore, the de Morgan law for disjunctions holds. The de Morgan law for conjunctions is demonstrated by a dual argument.

Finally, consider the defining property of the top $\top$, i. e., $\top \succcurlyeq x$ for every $x \in X$. It follows from the strong law that $f(\top) \preccurlyeq f(x)$. Since f is onto, we have $f(\top) \preccurlyeq y$ for every $y \in X$. Therefore, we obtain $f(\top) = \bot$. A dual argument verifies that $f(\bot) = \top$. We conclude that X is an Ockham algebra. ☐

Lemma 10.8. *If X is a distributive Ockham algebra with polarity $f : X \to X$, then $x \vee f(x) \succcurlyeq f(x) \wedge f^2(x)$.*

Proof. By direct computation,

$$[x \vee f(x)] \vee [f(x) \wedge f^2(x)] = [x \vee f(x)] \wedge [x \vee f(x) \vee f^2(x)] = x \vee f(x).$$

Therefore, we obtain $x \vee f(x) \succcurlyeq f(x) \wedge f^2(x)$. ☐

Lemma 10.9. *Suppose X is an Ockham algebra in the Berman class $B(m,n)$. Then its Berman skeleton $S_n(X)$ is an Ockham subalgebra of X in the Berman class $B(m,0)$.*

Proof. It is straightforward to verify that $\top$ and $\bot$ are elements of $S_n(X)$. Let $y_1 = f^n(x_1)$ and $y_2 = f^n(x_2)$. Thus, y_1 and y_2 are generic elements in $S_n(X)$. Since the polarity $f : X \to X$ satisfies the de Morgan laws, we obtain

$$f^k(x_1 \vee x_2) = \begin{cases} y_1 \wedge y_2 & \text{if } k \text{ is odd,} \\ y_1 \vee y_2 & \text{if } k \text{ is even} \end{cases}$$

and

$$f^k(x_1 \wedge x_2) = \begin{cases} y_1 \vee y_2 & \text{if } k \text{ is odd,} \\ y_1 \wedge y_2 & \text{if } k \text{ is even} \end{cases}$$

by induction on $k \geq 1$. In particular, these formulas hold when $k = n$. Furthermore, it is obvious that $f \circ f^n(x) = f^n \circ f(x)$. Thus, $y \in S_n(X)$ implies $f(y) \in S_n(X)$. We conclude that $S_n(X)$ is an Ockham subalgebra of X. Finally, suppose that $y = f^n(x)$, i. e., $y \in S_n(X)$. Then

$$f^{2m}(y) = f^{2m+n}(x) = f^n(x) = y.$$

Therefore, $S_n(X)$ belongs to the Berman class $B(m,0)$. ☐

Theorem 10.10. *Suppose X is an Ockham algebra in the Berman class $B(m,0)$. Then the polarity $f : X \to X$ is invertible and the strong law of polarity holds.*

Proof. For every $x \in X$, we obtain $x = f[f^{m-1}(x)]$. Thus, the polarity f is onto. Similarly, $f(x) = f(y)$ implies $f^{m-1}[f(x)] = f^{m-1}[f(y)]$. Thus, f is one-to-one. Therefore, f is invertible.

We next show that the inverse $f^{-1} : X \to X$ satisfies the de Morgan laws. Let $x = f(\hat{x})$ and $y = f(\hat{y})$. Since $f(\hat{x} \vee \hat{y}) = f(\hat{x}) \wedge f(\hat{y}) = x \wedge y$, we obtain $\hat{x} \vee \hat{y} = f^{-1}(x \wedge y)$. Thus, $f^{-1}(x) \vee f^{-1}(y) = f^{-1}(x \wedge y)$, which verifies the de Morgan law for conjunctions. The proof of the de Morgan law for disjunctions is similar.

Since f^{-1} obeys the de Morgan laws, it follows from Lemma 10.3 that f^{-1} is a polarity. Thus, $f(x) \succcurlyeq f(y)$ implies that $f^{-1}[f(y)] \preccurlyeq f^{-1}[f(x)]$, which reduces to $x \preccurlyeq y$. Therefore, the strong law of polarity holds. □

Lemma 10.11. *Suppose X belongs to the Berman class $B(m,0)$. Every element $x \in X$ belongs to a cycle, and the least common multiple $2k$ of the lengths of the cycles corresponds to the least parameter such that X belongs to $B(k,0)$. Furthermore, the parameter m is a multiple of k.*

Proof. For every $x \in X$, let $k(x) = \min\{j \geq 1 : f^j(x) = x\}$. Since $f^m = f^0$, we know that $k(x) \leq m$. Define $k_0 := \operatorname{lcm}\{k(x) : x \in X\}$. Since $\{\top, \bot\}$ is a cycle of length 2, we find that k_0 is an even number. Thus, we obtain $k_0 = 2k$ for some $k \geq 1$. Then k is smallest natural number such that $f^{2k} = f^0$. Thus, we obtain $m \geq k$. Dividing m by k, we obtain $m = qk + r$ from the Chinese remainder theorem, where integer q is the quotient and integer r is the remainder ($0 \leq r < k$). Notice that $f^{2m} = f^{2(qk+r)} = f^{2r} = f^0$. Therefore, we obtain $r = 0$ by virtue of the selection of k. We conclude that $m = qk$. □

Lemma 10.12. *Suppose X is an Ockham algebra of the Berman class $B(m,n)$. If the polarity $f : X \to X$ is one-to-one, then it is onto and thus invertible.*

Proof. Contrary to hypothesis, assume f is not onto. Thus, we obtain $X \neq S(X)$. It is easy to verify that

$$S_{k+1}(X) \subseteq S_k(X) \subseteq \cdots \subseteq S_1(X) = S(X)$$

for every $k \geq 1$. Select $x \notin S(X)$. Let $k = \min\{j : f^k(x) \in S_n(X)\}$. Then we obtain $k \geq 1$. Define $y := f^{k-1}(x)$. By construction, we obtain $y \notin S_n(X)$ and $f(y) \in S_n(X)$. Define $z := f^{2m-1}[f(y)]$, so that $f(z) = f(y)$. Since $z \in S_n(X)$ and $y \notin S_n(X)$, we must have $y \neq z$. Therefore, the polarity f is not one-to-one. This contradiction verifies the result. □

Theorem 10.13. *Suppose X belongs to the Berman class with parameters m and n, where m is an odd number. Then the map $f^m : S_n(X) \to S_n(X)$ is a well-defined polarity such that $S_n(X)$ is a de Morgan algebra with respect to f^m.*

Proof. For brevity, let $g := f^m$ restricted to the domain $S_n(X)$. Then for every $x \in S_n(X)$, we obtain $x = f^n(y)$ for some $y \in X$. Thus,

$$g(x) = f^m \circ f^n(y) = f^{m+n}(y) = f^n \circ g(y).$$

Thus, we find that $g(x) \in S_n(X)$. Since m is an odd number, $g : S_n(X) \to S_n(X)$ is a polarity. Furthermore, we obtain

$$g^2(x) = f^{2m}(x) = f^{2m+n}(y) = f^n(y) = x.$$

Therefore, the map g satisfies the law of double negation. Since f satisfies the de Morgan laws and m is odd, we obtain

$$g(x_1 \vee x_2) = f^m(x_1 \vee x_2) = f^m(x_1) \wedge f^m(x_2) = g(x_1) \wedge g(x_2)$$

and

$$g(x_1 \wedge x_2) = f^m(x_1 \wedge x_2) = f^m(x_1) \vee f^m(x_2) = g(x_1) \vee g(x_2).$$

Thus, the map g satisfies the de Morgan laws. □

Corollary. *If X belongs to the Berman class $B(m,n)$ with respect to the polarity f, where m is odd; then X belongs to the Berman class $B(1,k)$ with respect to the polarity f^m, where $k = \lceil n/m \rceil$.*

Proof. Assume m is odd, so that f^m is a polarity. Notice that $m\lceil n/m \rceil \geq n$. Since $f^{2m+n} = f^n$, we obtain $f^{2m+m\lceil n/m \rceil} = f^{m\lceil n/m \rceil}$ by raising the exponents. Rearranging exponents, we obtain $(f^m)^{2+\lceil n/m \rceil} = (f^m)^{\lceil n/m \rceil}$. The result immediately follows. □

10.5 Exercises

(1) Suppose $X = [0,1]$ is endowed with fuzzy logic. Let δ denote a real number such that $1/2 < \delta \leq 1$. Designate an element $x \in [0,1]$ as "true" provided that $x \geq \delta$. Verify that the following reasoning is sound: $p, \sim p \vee q \vdash q$.

(2) Suppose $X = [0,1]$ is endowed with fuzzy logic. Verify that $q < 1/2$ and $\max\{1 - p, q\} > 1/2$ imply $p < 1/2$. Explain why this inference is the contrapositive of the disjunctive syllogism in fuzzy logic.

(3) Suppose X is a de Morgan algebra. Let U denote an up-set of elements in X designated as truth. Thus, "x is true" provided that $x \in U$. Verify that the following categorical syllogism is sound: $p, q \succcurlyeq p \vdash q$.

(4) Suppose the lattice X is equipped with the polarity $f : X \to X$. (i) If $f^2 \preccurlyeq f^0$, demonstrate that $f^3 = f$. (ii) Construct a nontrivial example of a polarity such that $f^2 \preccurlyeq f^0$, but $f^2 \neq f^0$. (This confirms that $f^3 = f$ does not imply $f^2 \succcurlyeq f^0$.)

(5) Suppose X is a finite lattice. Demonstrate that pseudocomplementation defines a polarity $x \mapsto x^\star$ and that $x^{\star\star} \succcurlyeq x$.

(6) Suppose X is an Ockham algebra with polarity $f : X \to X$. Verify that the "negation" of $x \wedge f(x)$ is nonempty.

(7) Suppose X is an Ockham algebra with polarity $f : X \to X$. (i) Demonstrate that the map $f^2 : X \to X$ is an endomorphism, i. e., a homomorphism of X into itself. (ii) Further assume X is finite. Verify that f^2 is an automorphism if and only if f is one-to-one.

(8) Suppose X and Y are Ockham algebras. (i) Demonstrate that $X \times Y$ is an Ockham algebra. (ii) Further suppose X is the Berman class $B(m_1, n_1)$ and Y is the Berman class $B(m_2, n_2)$. Demonstrate that $X \times Y$ is the Berman class $B(m,n)$, where m is the lowest common multiple of $\{m_1, m_2\}$ and $n = \max\{n_1, n_2\}$.

(9) Suppose X is an Ockham algebra of the Berman class $B(m,0)$, where m is an odd number. Verify that every cycle of odd length is an antichain.

(10) Suppose X is a finite Ockham algebra with polarity $f : X \to X$. If f satisfies the strong law of polarity, demonstrate that (i) $\Phi \circ f(x) = 1 - \Phi(x)$ and that (ii) $\Phi \circ f^2(x) = \Phi(x)$.

Bibliography

[1] G. Birkhoff, *Lattice Theory*, Revised Edition, American Mathematical Society, 1948.
[2] G. Grätzer, *Lattice Theory: First Concepts and Distributive Lattices*, Dover Publications, 2009.
[3] G. Birkhoff, *Lattice Theory*, Revised Edition, American Mathematical Society, 1948, 139–40.
[4] T. S. Blyth, *Lattices and Ordered Algebraic Structures*, Springer Verlag, 2005, 52.
[5] T. S. Blyth and J. C. Varlet, *Ockham Algebras*, Oxford University Press, 1994, 6.
[6] T. S. Blyth and J. C. Varlet, *Ockham Algebras*, Oxford University Press, 1994, 4–5.
[7] T. S. Blyth, *Lattices and Ordered Algebraic Structures*, Springer Verlag, 2005, 55.
[8] N. D. Belnap, How a computer should think, in *New Essays on Belnap-Dunn Logic*, Cham: Springer International Publishing, 2020, 35–53.
[9] G. Grätzer, *Lattice Theory: First Concepts and Distributive Lattices*, Dover Publications, 2009, 72–3.
[10] L. A. Zadeh, Fuzzy sets, *Information and Control*, **8**, 1965, 338–53.
[11] P. Hájek, *Metamathematics of Fuzzy Logic*, Kluwer Academic publishers, 1998.
[12] Y. Shramko and H. Wansing, Some useful 16-valued logics: How a computer network should think, *Journal of Philosophical Logic*, **34**, 2005, 121–53.
[13] L. A. Zadeh, Fuzzy sets, *Information and Control*, **8**, 1965, 338–53.
[14] T. S. Blyth, *Lattices and Ordered Algebraic Structures*, Springer Verlag, 2005, 70–1.
[15] T. S. Blyth, *Lattices and Ordered Algebraic Structures*, Springer Verlag, 2005, 122.
[16] T. S. Blyth and J. C. Varlet, *Ockham Algebras*, Oxford University Press, 1994, 5.
[17] P. Geach and M. Black, *Translations from the Philosophical Writings of Gottlob Frege*, Basil Blackwood Oxford, 1960.
[18] G. Priest, Paraconsistency and dialetheism, in *Handbook of the History of Logic, Volume 8, The Many Valued and Nonmonotonic Turn in Logic*, edited by D. M. Gabbay and J. Gabbay, Elsevier 2007, 129–204.
[19] G. Priest, The logic of paradox, *Journal of Philosophical Logic*, **8**, 1979, 219–41.
[20] T. S. Blyth and J. C. Varlet, *Ockham Algebras*, Oxford University Press, 1994, 10.
[21] T. S. Blyth, *Lattices and Ordered Algebraic Structures*, Springer Verlag, 2005, 11.
[22] T. S. Blyth and J. C. Varlet, *Ockham Algebras*, Oxford University Press, 1994, 17–9.
[23] T. S. Blyth and J. C. Varlet, *Ockham Algebras*, Oxford University Press, 1994, 8.

https://doi.org/10.1515/9783112235812-011

Index

Antiatom 96
Antichain 22, 68, 71, 83–87, 104
Atom 32, 58, 96–100
Automorphism 7, 46–50, 129

Belnap information lattice 35, 48, 60
Belnap logical lattice 35, 60, 104–105, 118, 119
Belnap representation scheme 34–35
Berman class 126–129
Berman skeleton 126–129
Birkhoff criterion 25
Birkhoff representation theorem 35, 66
Boolean algebra 12–13, 14, 37, 58–60, 84–85, 86, 97, 124
Boolean center 14
Boolean hinge 98–103
Boolean truth lattice 37, 42, 59, 86, 94, 98–103
Bottom 9, 95
Bounded lattice 9

Canonical median 19, 22, 45, 83
Canonical score 18–19, 44–45, 57
Categorical syllogism 116
Chain 9–10, 19–20
Codual order 22
Complement 12–13
Congruence 95
Conjunction 7
Conjunctive syllogism 116
Contrapositive 117
Convex set 70–71, 72, 78–80

Dedekind criterion 10, 13–14
De Morgan algebra 12, 32, 36–37, 41–42, 46, 79, 94, 95–96, 98, 104–105, 116–120, 125, 128
De Morgan laws 11–12, 122–123, 125–126
Diamond equation 20–21
Diamond isomorphism theorem 11
Diamond lattice 10–11, 31–32, 35
Discrete fuzzy logic 56–58, 60–62, 96–98
Disjunction 7–8
Disjunctive syllogism 116
Distributive law 8, 12–14, 24–25, 67, 69, 71–72, 100, 104, 118, 124–125, 126
Double negation 11–12, 122–124
Down-set 18, 37, 44–45, 66–69, 70–72, 78, 81–82, 98–99, 102–103

Dual automorphism 12, 122, 129

Emptiness 14–15, 31–38, 41–44, 46, 56, 57–58, 61–62, 78–83, 95, 99–101, 119, 126
Equivalence relation 9, 22, 23–24, 95
Explosive logic 119

Filter 69, 82, 99, 103
Finitary convex set 93 (71–72)
Finitary down-set 68–69
Finitary up-set 69
Fixed point 46–49
Fuzzy logic 56–58, 81, 86, 96–98, 99, 120–122
Fuzzy set 60–62, 86
Fuzzy truth lattice 56

Hasse diagram 9–10
Heyting algebra 123–124
Homomorphism 8, 95–99, 102–105

Ideal 66–69, 82, 99, 102–103
Idempotent 47–48, 129
Implication 12, 116–122
Injective 7
Interval 10, 24, 37–38, 47–48, 56, 60–61, 70–72
Invariant subset 46–47
Involution 11–15, 32, 37, 41–42, 57, 61, 66–67, 71, 118–119, 122–123
Isotone 97
Isomorphism 7, 10, 11, 25, 37, 42, 57, 59–60, 62, 80, 84

Jordan–Dedekind chain condition 19, 21

Kleene algebra 80–83

Lattice 7–8, 41, 66–69
Lattice order 8–11, 18–22, 44–46
Left-right symmetry 46–50, 129
Logic of paradox 45, 57–58, 61–62, 96–98

Material implication 117–122
Maximal element 66, 68–69, 71–72
Middle region 78–80, 83–87
Minimal element 69, 71–72
Modular law 8, 10–11, 19–23, 46
Monomorphism 7

https://doi.org/10.1515/9783112235812-012

MS-algebra 125

Nagarjuna's identity 14–15, 31–32,126
Negative transitivity 22–25, 41, 81, 84
Nonemptiness 15–16, 31–33, 36–38, 41–44, 56, 79–80, 81–83, 86, 95, 99–101, 119–121

Ockham algebra 124–129
Oneness 14–15, 32, 126
One-to-one 7, 46, 122–124, 125, 128
Onto 7, 46, 124, 126
Ortholattice 13–14, 103

Paraconsistent 119
Partial order 8
Pentagonal lattice 10–11, 14
Pentalemma 24, 67
Polarity 11–12, 118, 122–126
Power set 12–13, 34–35, 58–60, 118
Predecessor 9
Principal down-set 18, 37, 66–69, 71–72, 81–82, 98–103
Principal up-set 18, 37, 69–72, 81–82, 98–103
Product algebra 42–44, 48–50, 58–62
Product order 43, 44–46, 61
Pseudocomplement 124–125

Simple truth lattice 104–105
Skeleton 122–125
Standard median 21–22
Standard score 19–22, 46, 57, 61
Subalgebra 7, 14–15, 37, 44, 46–49, 56, 80, 95–96, 124, 127, 129
Successor 9–10, 96
Surjective 7, 97, 122, 128

Tetralemma 15, 31–38, 41–44, 48–49, 59–60, 70–71, 78–79, 99, 101–104, 126
Top 9, 95
Truth function 18, 56–57, 61
Truth lattice 31, 56, 94–105, 119
Truth-preserving 95, 104
Truth-value 31–35, 56, 61, 79, 94, 100, 104–105, 118–122

Up-set 18, 37, 44–45, 69–72, 78–79, 81–82, 98–99, 102–103, 119

Printed and bound by CPI Group (UK) Ltd, Croydon, CR0 4YY
16/09/2026
14963228-0001